WTF Is Happening

W

Women
Tech
Founders
On The Rise

T
F

is Happening?

NISA AMOILS

LIONCREST
PUBLISHING

WTF IS HAPPENING

Women Tech Founders on the Rise

ISBN   978-1-5445-0289-2 *Paperback*

978-1-5445-0290-8 *Ebook*

This book is dedicated to my family.

Contents

Foreword

By Craig Enenstein, CEO, Corridor Capital

I am a mainstream contrarian. I always have been. Throughout my life, I have gone and sat at the proverbial lunch table with the new kid, the exchange student, the enigma. In high school, rather than join one of the popular fraternity-like clubs on campus, I formed the club with only one rule—all were welcome. We were an eclectic bunch, but we sure had fun. I learned that much of power in the social world comes from exclusionary behavior, to exclude others based on family name, appearance, wealth, academic alma mater, unwillingness to conform, gender. As a contrarian, my solution was inclusivity.

My very first experience at Wharton preterm was a retreat to the Poconos Mountains, where we started off in a room and mingled. On hearing we should all freeze, an instructor pointed out a white line down the middle of the room that divided the group roughly evenly. We were told to form groups of three with those on our side of the line. We were then told to join our three-person group with another from the other side of the line. This six-person formation was to be our Learning Team, tracking across all core classes for the first year of our two-year program.

As someone getting a master's in international studies from the Lauder Institute at the University of Pennsylvania along with my Wharton MBA, I intuitively selected for diversity and paired with students from Japan, Argentina, Hong Kong, and Taiwan. The one other American on my team was also in a dual master's program, studying international relations through Johns Hopkins. We were two women and four men, a reflection of our class, which had a similar ratio. On looking around the room, I noticed some groups of six white men, all with Wall Street experience. They had big smiles, as if they knew a joke that I didn't.

I learned over that year the difficulty of every group writing assignment, since only two of us grew up in the US university system. However, our Japanese student had been an investment banker and was great with modeling and finance; the Argentine had been a brand manager for a leading consumer products company and had marketing down cold. Experiencing diversity through collaboration on a small leaderless team turned out to be challenging and marvelous.

From my first semester of business school, I set my sights on private equity. In researching the industry, I learned there to be three drivers of value creation in the field: buy low, sell high (multiple arbitrage); the leverage in the buyout (application of leverage-enhancing equity returns); and the change in earnings over the holding period. The contrarian in me believed items one and two: the financial engineering that had traditionally made the asset class so lucrative would commoditize as great returns and would bring an excess of capital to the sector. I wanted to become The Earnings Guy, to be able to bend the earnings curve to achieve superior risk-adjusted returns.

Years later, in launching a private equity firm, we set out to apply investment, strategy, and operations experience to strong industrial and commercial lower middle market companies that are at inflection points such that active support for infrastructure enhancement, organic growth, and acquisitions would change the attractiveness of the underlying companies and deliver exceptional returns to management partners and our investors. To be able to purchase control of entrepreneur-run companies with the objective of teaming with management to transform businesses required leaning on my life experiences of inclusivity and collaboration.

Classically, private equity was popularized by well-documented public takeovers and strong personalities, sometimes referred to as Masters of the Universe. While being acquired by private equity was considered quite attractive for a time, the bloom is off the rose, as many stories have emerged about the often unpleasant, sometimes arrogant behavior of private equity investors toward management teams. Over time, we developed an approach we consider to be EQ (Emotional Quotient) equal to IQ (Intelligence Quotient), something we learned to be less common in the private equity field. We developed a style we occasionally refer to as "Servant Capital," through which we recognize the need to look beyond the simplicity of control dynamics and, more importantly, "serve" our portfolio company management teams. This is a warmer, more collaborative approach that we have found to be rather attractive to entrepreneurs and management teams.

Along the way, without design, we found that our approach was exceptionally compelling for women business leaders who have shared their appreciation for a more respectful, collaborative style that includes active listening, regular feedback, and transparency. Consequentially, since 2011, women have represented

38 percent of our platform CEOs, and beyond CEO, 50 percent of our company's C-suites have included women executives. I have sadly heard too many women share stories of difficulty accessing capital or having poor experiences with a male-dominated investor world.

Women CEOs represent 17.5 percent of privately held businesses with paid employees. For the leading fifteen hundred S&P companies, the figure drops to a dismal 5.1 percent. So who are these women who run companies? I believe they are the best of the best. To achieve such a level of responsibility in a system that has long been unwelcoming takes extraordinary skill and fortitude. This leadership population is resilient, ingenious, intelligent, and driven. Yes, I am stereotyping these women CEOs and recognize there are clearly exceptions. We are thrilled to back such talented individuals. The CEO of a company into which we recently invested flattered us during the announcement to the employees that the company's first outside investor, us, was selected based on emotional intelligence.

Over the past thirty years, I have had the privilege of knowing phenomenal women in business. At the Wharton School of the University of Pennsylvania, 30 percent of my class was women, an exceptional group. More recently, through participation in Young Presidents Organization (YPO), the leading global CEO organization, I have had the honor of getting to know women leaders from around the world. These interactions have only increased the strength of my conviction in the caliber of the talent pool of women executives. I am forever indebted to the education I have received from these great women.

I will continue making investments in talented women in business. It is the economically sensible choice. Remember, I am passionate about quality risk-adjusted returns, and I see favorable economics in engaging women in leadership roles. I predict that this era of inefficiency where women are underfunded, overlooked, or otherwise not treated with the dignity all people seek will come to an end. Over the long run, markets are efficient. Smart money ultimately goes to quality leaders.

Introduction

In late 2018, I visited the offices of one of the largest block-chain/cryptocurrency venture capital firms. During the course of a great discussion exchanging company information, I asked one of the managing partners if they had any female-founded companies in their portfolio.

"I don't think that exists in this sector," he replied. Between my associations with Women in Blockchain, Women on the Block, and my general network, I could produce a long list of female-led blockchain companies off the top of my head. I said, "Yes, it does exist—they just don't know how to find you, and you don't know how to find them."

That's exactly the moment I knew I had to write this book. After reading Emily Chang's *Breaking up Brotopia*, about the male-dominated technology industry in Silicon Valley, I knew someone needed to pick up where she left off with these newer, emerging technologies. Since they are newer, the cycle needs to be broken now, before the culture is ingrained, and women need to play a major role in shaping them.

The Gender Funding Discrepancy

Throughout my career as a technology investor, I have watched female tech entrepreneurs struggle for venture capital (VC) funding as the money flows elsewhere. In 2018, female founders still received only 2 percent of VC funding (when there is no male cofounder—that number goes up to around 13 percent when there is a male cofounder).*

There have been many studies that show that women entrepreneurs outperform. To take one frequently cited example, First Round Capital did a thorough review of their own portfolio in 2015 and discovered that companies with at least one female founder did a better job creating value for investors. A much better job, in fact: those women-founded companies outperformed companies started by all-male teams by 63 percent. A 2013 study com-

---

* Laura Huang, "Why Female Entrepreneurs Have a Harder Time Raising Venture Capital," *Scientific American*, June 5, 2018, https://www.scientificamerican.com/article/why-female-entrepreneurs-have-a-harder-time-raising-venture-capital1/.

missioned by the Small Business Administration came to the conclusion that venture capital firms that invested in women-led businesses (WLBs) saw an improvement in their firm's performance. The paper was unequivocal in its conclusions both that the potential of women-owned companies is being held back by the lack of investment, and that those very companies are excellent investment opportunities: "WLBs may hold untapped potential for innovation, job creation, and other economic contributions that may be limited only by access to VC funding... investments in WLBs are successful, leading to a positive return on the VC firm investments."

Calvert Impact Capital, a nonprofit that invests in companies with environmental, social, and governance goals (ESG), recently released a study, which included eleven years of data on 160 global companies that it holds in its portfolio and represent $23 billion in assets, called "Just Good Investing: Why Gender Matters to Your Portfolio and What You Can Do about It." Commenting on the study, Leigh Moran, director of Calvert's strategy team, said, "People think of investing through a gender lens as a niche, part of impact investing, but it's just a part of good investing. This study says if you want to maximize returns, seek out companies with gender diversity."

The study found:

- The top quartile for gender diversity in management had a return on sales of 18.1 percent and a return on assets of 3.9 percent; the bottom quartile had a return on sales -1.9 percent and a return on assets 0.3 percent. The return on equity was 8.6 percent versus 4.4 percent.
- Companies where women were more than 57 percent of senior management teams were in the top quartile; those with 20 percent or less were in the bottom quartile.
- Once women represent 33 percent of leadership, increase in financial performance is more significant; it's not just the numbers of women, but the ratio of women to men that matters.

With all this data showing that female founders outperform, why does this funding discrepancy persist? There are many cited reasons. One of the most frequently cited is the lack of female investors to whom the deals are sourced. In 2018, women made up 8 percent of investing partners at US VC firms. A whopping 74 percent of US VC firms have no women. The industry is staggeringly homogeneous and operates on pattern matching. Under pressure from the press and industry associations, there have been efforts in recent years to place token female investors at large firms, and many more launches of "gender lens"-focused micro (smaller asset) VCs and handful over $100 million AUM, more angel investors, crowdfunding, and other sources of capital for female founders.

We now have a Billion Dollar Fund, which is a consortium of VCs geared toward this goal. However, this level of investment represents a disproportionately small fraction of available institutional and retail capital. According to a 2017 study by the Knight Foundation, only 1.1 percent of the total assets under management is with firms owned by women. The problem goes beyond VC up the capital stack. As companies scale into growth and private equity, the problem of unequal funding persists and so does the low percentage of female investors. This is where the pensions, foundations, endowments, sovereign wealth funds, and other entities that fund the funds need to enact macro reallocation of capital. I am on the Advisory Board of Girls Who Invest and also part of a lobbying group of VCs who are working to change the structural barriers there.

As of early 2019, according to All Raise, a nonprofit trying to diversify the VC industry (of which I am also part), the number of women in check-writing roles has seen a bump, but it's not much. American VC firms have added thirty new female senior investing partners in the last ten months, according to All Raise, which analyzed publicly shared hiring news. Firms hired more women in the second quarter of 2018 than any quarter in at least the previous five years. That has felt like a lot to observers. But during the same time, the industry added sixty-eight male partners. So the percentage of women in leadership roles only rose to 9.5 percent from 8.9 percent. About three-quarters of US venture capital firms still have zero women partners. Among the 153 firms that do boast a woman partner, it's often singular—not plural: three-quarters of these firms only have one woman in a senior leadership position. Those are roughly the same figures as ten months earlier, according to All Raise. The core of the problem: VC firms typically recruit from the C-suites of big tech companies (mostly men), from the entrepreneurs whom they've previously backed (mostly men), or from other venture capital firms (mostly men). So recruiting more women often means changing the way firms recruit, another barrier.

Even if there were many more female investors, that is no silver-bullet solution, as bias against female founders would not automatically disappear, and female founders should not be restricted to working only with female investors. A *Harvard Business Review* study from 2017 analyzed video transcriptions of pitch sessions and found that male and female entrepreneurs are asked different questions by VCs, which affects how much funding they get. For instance, VCs tended to ask men promotion questions about hopes, achievements, advancement, and ideals. With women, they focused on prevention questions about safety, responsibility, security, and vigilance. The companies with promotional questions raised more capital.

Another reason the funding gap persists is that for many investors, it may require more work to find female founders. They generally seek companies to fund from their own networks, companies that tend to be founded by people similar to themselves. Pattern matching is mistaken for opportunistic investing. This approach will continue to be challenged by clear demographic shifts in the US and globally. According to *Harvard Business Review*'s study this summer, "Finally, Evidence that Diversity Improves Financial Performance," homogeneity imposes financial costs as decision making suffers.

The good news is we are on the way, even though change is slow. Fortunately, according to a report from Veris Wealth Partners, over the past year (2018), investments to benefit women surged to $2.4 billion from $100 million four years earlier, a twenty-four-fold increase. Female founders who have taken their companies public are helping to re-create the funding ecosystem by setting up funds or investing in others.

Wouldn't it be great to accelerate the pace of change? With all the cutting-edge technology accelerating now, it is important that women take part in shaping the future. The often-cited excuses, that women don't have enough mentors and role models and are operating from the mentality of scarcity and so are not supporting each other, are fading with the new, bold generations. These women are on the cutting edge—they need to have the light shone on them, so that other women can follow.

A Future the Jetsons Would Recognize

Women are involved in amazing things happening in technology right now. Here are some examples:

- We are closer to autonomous flight than most people comprehend. Vertical Takeoff and Landing (VTOL) technology is in rapid development, leading the way to autonomous air taxis (you might be riding in a driverless aerial cab in a decade).
- Drones are rapidly expanding their roles in society: conducting security checks, working in agricultural fields, going into mines, delivering packages—anything dirty, dull, or dangerous. Some drones are using LIDAR and sensors to fly autonomously.
- The intersection of robotics with artificial intelligence (AI) is transforming manufacturing and some consumer tasks. Social robots, which help care for the elderly and the ill by keeping them company, managing their medicines, and watching out for them,

are finding a growing number of uses, particularly in countries that embrace them, such as Japan.

- In the virtual and augmented reality (VR/AR) world, innovations can be used to teach history in an immersive, visceral, and deeply memorably way. In business, VR/AR can bring together constituents on complex construction projects to experience designs before they are built.

There were so many female candidates to include in this book that I had a hard time narrowing it down, so I focused on areas in my own investing background: virtual and augmented reality, drones and robotics, artificial intelligence and machine learning, blockchain, and cryptocurrency (often called "frontier tech" or "hard tech"). I intentionally wanted to show women in these fields as these areas have fewer female founders than fashion, beauty, and ecommerce companies, which are often stereotyped as women's companies. While there are great founders in those and other areas, it is important that women are seen as mastering these hard tech areas and shaping the future.

By no means is this book a complete representation of the full scope of diversity among female founders. Rather, it's a snapshot of a few of the many impressive women I know of. (And frankly, there is so much innovation happening in blockchain, cryptocurrency, and digitization of assets that I could have filled these pages solely with those founders—so you'll find a lot of them here.) I only focused on gender and left other kinds of diversity, such as race, out of this book—that could be another book entirely.

In the following pages, you will find a dozen Q&A interviews with inspiring female entrepreneurs. I asked each of them to share their experiences in their own words. This book is not a promotion for the companies included here. Nor is it a rundown of my investments—I am not an investor in any of the firms represented, and this should not be construed as offering investment advice. Finally, you will not find many hard numbers about returns, because it is simply too early for these founders to say how things are working out. The women I profiled are in the thick of building their companies, with exits somewhere down the line. Their technologies are young, their industries are young, but they are on the right path. I am confident that these women, and women like them, will outperform for their investors.

Your Next Big Investment

Where will alpha returns come from in the future? Aside from the well-published social benefits to society and moral imperative arguments, there are real investment opportunities here. These women-led companies are often undershopped and undervalued. Investors who understand this have the opportunity to make a higher return by investing (it's classic arbitrage). Investors who aren't looking hard are leaving money on the table and, eventually, will be left behind.

This book is really an invitation. It's an invitation to young women who may not have thought about working in STEM fields (Science, Technology, Engineering, and Math, sometimes also called STEAM, with Arts), or starting their own tech companies. It's an invitation to female entrepreneurs and investors to start more companies. Most importantly, this book is an invitation to investors who are seeking alpha returns and outperforming companies, inviting them to see the amazing world female technologists are creating. I invite you to join in...

Dawn Song / Oasis Labs

Dawn Song is CEO and cofounder of startup Oasis Labs, a privacy-first, cloud computing platform on blockchain. A professor in the Department of Electrical Engineering and Computer Science at the University of California at Berkeley, she is an expert on security and privacy technologies for software, networking, distributed systems, applied cryptography, blockchains, smart contracts, and machine learning. Named a MacArthur Fellow in 2010 at age thirty-five, she has also been awarded the Guggenheim Fellowship, the Alfred P. Sloan Research Fellowship, the NSF CAREER Award, the MIT Technology Review TR-35 Award, as well as the Faculty Research Award from IBM, Google, and other major tech companies. She is an accomplished entrepreneur responsible for multiple startups, including Ensighta Security (acquired by FireEye Inc.) and Menlo Security. She obtained her PhD from UC Berkeley.

Why did you start your company?

I've been doing research on blockchain for quite some time, and while we see that on one hand blockchain has the opportunity to really help and transform many industries, it is still at a very early stage. Right now, there are serious scalability issues and no real privacy protection.

The Oasis Labs team has been developing technologies that we hope will build new solutions for exactly these issues. Our mission is to bring blockchain technology to a new level, by providing a new computing paradigm and platform to enable applications to be built that couldn't be built before. What Oasis Labs comes down to is recognizing that there is a serious need within real-world applications and that we have something that can help solve these issues and change the world.

Add to that, we all hear news every day of the internet's significant and pressing privacy issues. The media talks about Cambridge Analytica, data leaks, Equifax, privacy breaches—it's everywhere, and no one can escape it. All of these issues leave me with the sense that, unless we do something different, these problems will only become worse. As a researcher and leader in this field, I feel that I have a responsibility to build new technologies that can help solve these problems, and I am glad to see it coming together in the start of Oasis Labs.

Where did your interest in building companies and products come from?

Oasis Labs isn't my first company. I've started two companies previously. My first startup (which was acquired) was in mobile security, and the other was a web security company for secure browsing. That web security company now has its own dedicated team and is very successful. All my companies share a common theme: they incorporate technologies that were developed in my research group at UC Berkeley. So I have been leading and developing the research innovation with my students and collaborators for all of these companies from the very beginning. Then each one has gone through the process of developing and deploying the technology into a real-world, working product. They all have an underlying theme of providing a technology that aims to improve and innovate on specific functions and have a real impact on the industry they serve.

Different researchers have different preferences, but for me, I'm very excited about building new technologies that, on one hand, have real innovation and creative ideas and, on the other hand, can help make the world a better place. To me, that's a huge part of our responsibility as researchers—we don't just do research in isolation; instead, we use our research to help improve the world.

Plus, from a purely scientific sense, deploying our new innovations is a great way to validate our ideas and technologies. We can then learn how to improve upon our approach from lessons learned in real-world deployment.

What advice would you give to other women in your field?

I've always been in the gender minority—in high school (I was in a special class assembled to train for Math Olympics), in undergrad as a physics major, in graduate school in my computer science PhD program, etc. My advice to women is to be confident. Believe that you can achieve all the things that anyone else can achieve. As female leaders, we tend to doubt ourselves and worry what others will think. We need to stop worrying what others will think and instead lead with conviction.

What advice would you give to STEM students?

Learn basic science to have a solid foundation. When I was in school, I started as a physics major in undergrad but then switched to computer science in graduate school. Once you have a solid foundation, it's much easier to move to other domains and be adaptive.

Also, keep a curious mind. Be brave and explore new and uncharted domains. I'm unique in having a much broader area of research than most people. For example, I've worked in computer security (including most subfields in security), AI and machine learning, and of course, blockchain. I have really enjoyed exploring new questions and new domains, and in my career, that has been incredibly helpful. I'm always curious and excited to explore uncharted territory, and as a result, I've been one of the earliest to explore many of the domains I've worked in.

What do you predict will happen in your field in the next five years?

I am fortunate to work in several important and exciting fields: security and privacy, AI and machine learning, and blockchain. All of these fields are very fast moving. One trend I think we will continue to see is that technology advancement will keep getting faster and faster. And it's happening across all different domains.

In general I think there will be tremendous technical advancement and change in all my domains and industries, but because everything is changing so quickly, it's hard to predict where we will be even in five years. It's analogous to the beginning of the internet: we couldn't really imagine how people use the internet today.

I do think that what Oasis Labs is building will lead to significant changes in the world in five years. I think we can build a highly scalable platform with privacy protection that works and solves a lot of current issues, and I think we will see new types of applications that we can't even think up today.

What was your big breakthrough?

My big breakthrough was the Oasis Platform, which combines a number of new technologies that we have developed and are continuing to develop for practical privacy protection as well as scalability. This set of technical innovations enables our key technology on privacy-preserving smart contracts at scale.

Why should younger women consider STEM and/or entrepreneurship as a career option?

It's important to give people options. For women who want to go into STEM, we want to make it so that they have the option to do so. But that doesn't mean that every woman has to pursue STEM. No one is a failure if they don't go into STEM, and women shouldn't feel pressured to have to go into STEM. If a woman wants to be a full-time mom, that's great. If she wants to pursue a career in STEM, that's great, too. At the end of the day, as a community, what is important is to provide the option and equal opportunity for women and minorities so that anyone can join STEM.

I've always felt that working in STEM is really fun—it's fascinating, and I enjoy it. Working in STEM allows me to deepen my understanding of the world.

Similarly, entrepreneurship is a really fun, fascinating

experience, beyond the technical and scientific experiences. You are faced with challenges you have to tackle in widely different areas. Again, I'm always growth minded, and entrepreneurship brings me some of the best opportunities to grow, as a leader and as a person. It's difficult, but it's an incredible experience that can deeply enrich your life and, at the end of the day, help you become a better person!

Is there anything else you'd like people to know?

As a leader, one of my strengths is providing a technological vision as well as pushing forward technical innovation. I see a leader as someone who needs to inspire people toward the same mission to help make the world a better place.

I'm also very growth minded. I have a natural tendency to see the best in people and believe in that best part of them. Being a leader is similar to being a professor or an advisor. It's a leader's responsibility to help the people he or she leads to grow and to reach their full potential.

Carrie Shaw / Embodied Labs

Carrie Shaw works at the intersection of health education and virtual reality storytelling. She is the founder and CEO of Embodied Labs, an immersive education and wellness platform for professionals, families, and caregivers and the elders they serve. After graduating from UNC Chapel Hill with a BS in public health, she spent two years working as a Health Education Peace Corps volunteer in the Dominican Republic, where she came to understand the way visual communication tools have the unique potential to cross cultural, language, and education barriers. After returning to the US, she worked as a medical visualization research assistant and a teaching assistant in anatomy and physiology, while simultaneously serving as her mother's caregiver. Her mother's diagnosis of early onset Alzheimer's disease made her aware of the needs of caregivers and the aging services workforce, leading her to complete an MS in biomedical visualization and found her company. Embodied Labs' work has been featured by the AARP, United Healthcare, *Forbes*, *JAMA*, and The History Channel. In July 2019, Embodied Labs won Grand Prize and $250,000 from the Bill & Melinda Gates Foundation's XR in Education Prize Challenge.

What is your background?

When I was in my late teens, my mom was diagnosed with early onset Alzheimer's disease, and that has become a big part of what I'm doing today. When she was diagnosed, I was just starting college at the University of North Carolina, Chapel Hill, and I spent a lot of time being part of her care team. I studied public health and wound up being really interested in global health education.

Some of my first work was abroad, as a community health educator with the Peace Corps. I had to teach reproductive health and family nutrition, and my inability to speak the language and really understand the culture made it challenging to communicate information. That was where I first saw how visuals and storytelling can help cross some of those barriers. I actually didn't know until my midtwenties that medical illustration could be a career, but as I learned how art and visuals could be combined with science to teach human health in a clear and impactful way, I decided that was where I wanted to go next with my education.

In between all of that, I spent several years living with my mom as her primary caregiver, and I saw [the potential of medical visualization] firsthand as part of her care team. I had hired home health aides to work with her, but we all struggled to understand my mom's perspective and what she was going through. Her brain was changing. And that's when I asked this question: why can't we use what we know about different health conditions and experiences to actually see through someone else's eyes to better understand? I wound up making that part of my graduate research thesis for medical visualization at the University of Illinois at Chicago.

While I was prototyping answers to that question, I had no preconceived ideas of how I was going to answer it. I wound up moving through a few different prototypes and finding virtual reality. When I first tried one of the early virtual reality headset prototypes, I thought, *Here's a way we can actually re-create worlds and tell stories through a perspective different from our own.* I met some of the people who are now my business partners today in that prototyping process back in 2015. We started making a workflow that let us integrate live-action films with computer-generated gaming objects and then use a person's own hands projected as someone else's as a controller to give them agency as they navigated through what we call an embodied experience.

Why did you start your company?

I would have never guessed that I would be a startup founder. I really wanted to continue on the academic track, research forever, and be a professor. I actually had accepted a PhD offer to continue studying the theory of embodied cognition in immersive environments and how the things you do in an immersive environment can translate into skills in the workforce.

What took me away from that was that I started forming relationships with the people who are now my cofounding team. They had really different skills than mine, and we realized that if we combined our skills, we could innovate and build something new, which excited us. I saw that if we went all in, if we put our time and heads together not only to develop the technology but also potentially to create a business around it, taking what we've learned and actually applying it to an industry model, then we could get this technology out into the world more quickly and move it forward more than I could as one person in a PhD program. Working by myself would keep the full vision from being realized.

So, we really started thinking about the company in May 2016. By the end of the summer, early August 2016, we had incorporated officially as an entity. At first, I deferred my PhD, and then I eventually decided that it wasn't the right time to pursue it. It had become clear that there was huge potential, albeit also risk, for all of us to put our previous plans on hold and jump into forming this startup.

Where does your interest in building companies and products come from?

It comes from wanting to innovate faster. There are advantages to both academia and industry, and each has its own role in bringing about innovation. I like that in academia you can test out ideas that are high risk and maybe too early to be brought to market and turned into a full-fledged business. It's also a great place to test pieces of questions and hash out specific, more detail-oriented theories. Over time, even though I was really interested in academia, I became more and more frustrated with the pace of innovation. You're limited in some senses by the scientific method and the need to publish and peer review.

In doing my medical illustration degree, for the first time I was around creators who were going to build freelance careers making art and animation and lots of things that would immediately be turned around and put to use. What industry does really well is pull together different ideas, prototypes, people, and skill sets and combine them to create full-fledged products. I made the decision that I wanted to move into industry so that I could innovate at a faster pace: I could build a team with different skills, and then we could develop and actually see a product in the market and in the hands of end users.

How did you get your funding at each stage?

It's been a long and winding road. There hasn't been a magic button or easy access to capital. There never really is. I think being a woman and first-time entrepreneur has made it especially hard. Over the past two years, we've really had to prove our value and bootstrap more than we initially anticipated.

Our very first funding—$50,000 dollars of seed funding—came from my immediate family. One of the cofounders is my older sister, who has a background in education technology and curriculum, which has been really important to our product development. Having a couple of us from the same family on board with the company made it possible for us to approach our family and really share our vision with them.

From there, right after we incorporated, we went into a business accelerator program, a really small one. It was the first cohort that had ever been done in a program called Creative Startups in my hometown of Winston-Salem, North Carolina. We did that program because we honestly had no idea how to run a business. We had an idea for a product but had never been a part of the business world or any kind of startup prior to this. We actually wound up winning a $25,000 no-interest loan from that accelerator program. We also made some connections with angel investors who were part of the mentorship that the program provided, and they gave us our first investments from non-friends and family.

We didn't have a product on the market, and we had no paying contracts, though we did have some paid pilot customers come on in January 2017 in some of our initial markets. Chicago Methodist Senior Services was interested in buying our product and also saw potential in it and wanted to invest. Chicago Methodist Senior Services is a small nonprofit in Chicago that serves Continuing Care Retirement Communities (CCRCs), meaning they serve elders across the entire spectrum of needs. They invested some capital into Embodied Labs and became a customer-partner. That was important not only from a financial standpoint but also because it gave us a place to build our product and test immediately, allowing us to rapidly prototype and understand our market.

At the same time, we raised a small round of angel capital. We also won some different awards that gave us nondilutive cash: we won something like a $10,000 award for a medical empathy hackathon, and we won $50,000 from the US Department of Education after applying for their VR/AR educational simulation challenge.

All of that small preseed investment gave us what we needed to build a product that could get to market. Still, at that point we were bootstrapping and hadn't reached any kind of scale that would sustain a company or even prove that we had product-market fit. So from January 2017 to January 2018, we continued to build out more of our VR content library and realized we needed a better platform for distribution of our content. We also figured out that we needed to build in a learning framework and a data and analytics tool that would give people an idea of what is being accomplished through the VR-based experiences, and then how that ties back into budget and bottom line for our customers.

Initially, elder care was just one part of a broader picture, but we decided we should target where we had the most traction, which was in senior living facilities, home health, and home care. So we really focused in on elder care and set out to build a product for caregivers working with seniors.

In February 2018, we hit a big milestone when we won Most Viable Solution in a competition hosted by the AARP and United Healthcare specifically looking at solutions for caregivers of people living with dementia. That award came with some more nondilutive cash and also gave us credibility among our customer market, so that more people and organizations wanted to invest in our solution and become customers. That then enabled us to raise another round of capital from angels specifically in the aging care and technology market. Concurrently to these angel investments, in July 2018, we won $250,000 from the Bill & Melinda Gates Foundation as part of their challenge for workforce solutions using augmentative virtual reality.

Today we've just started to have product-market fit in our target market and some significant traction with larger-scale customers, so we're focused on being successful with those customers. We plan on raising our first institutional [funding] sometime in the next several months as we grow.

What advice would you give to other women in your field?

Just don't quit. It sounds really simple, but someone gave me that advice early on, and it's stuck. This advice is really important for women because there are way more obstacles [for women than men] if you're starting a venture-funded business, especially in technology.

I've thought about quitting at a lot of points, but even when it felt like we were hitting dead ends and weren't able to validate products and get funding fast enough or bring on customers, I would tell myself, "Well, I just won't quit." We just kept problem-solving and kept going. In the last three months, we've hit a place where I know it'll still be really hard, but I am glad I didn't quit, because I would have missed out on the success that we're finally starting to see.

Also, have confidence in yourself. Don't be afraid to be authentically yourself, and don't feel a pressure to lead in a certain way based on what you see out there. I've really tried to embrace my own leadership style and to understand that leadership can look a lot of different ways. I think there are fewer female role models doing what I'm doing, so I would advise women to be their authentic selves and see themselves as leaders from the get-go.

What advice would you give to STEM students?

Care less about grades and more about learning how to work with other people who are in different disciplines or who have different qualities than you. Actually try to build things—build out an idea into something tangible. I think classroom time is less important than being able to take creative risks, and school is a great time to try things out. When I was in graduate school, it was a safe place for me to ask and pursue whatever question I wanted to answer and to imagine something that didn't exist yet. I had no idea where this question would take me, but finding people to work with and build it out was where everything got started.

I'd also say, if something you're building or an idea you have seems silly or impossible, then take that as a sign of success, and pursue it if it is of interest to you. That's where innovation comes from.

Also, look at opportunities to travel or do internships and get life experience that puts you in places you haven't been before. Even now, I'm surprised when I look back at how many lessons I learned by being in the Peace Corps and by taking a couple of research abroad opportunities while in the public health program at UNC Chapel Hill. The problem-solving I had to do in these sorts of applied scenarios taught me a lot about business, how to work on my feet, and how to work with and understand lots of personalities and cultures. To the extent possible, get hands-on, applied experience in whatever field you're studying.

What do you predict will happen in your field in the next five years?

For virtual and augmented reality, I think we'll see those solutions merge into immersive technology, and we'll start thinking of it as its own media format. Instead of thinking about it as kind of like a movie and kind of like a game, it will just be immersive technologies. Kind of like the radio was or the TV or the internet, it'll become a major media platform and format that impacts all aspects of life in ways we're only just imagining right now, and in ways we haven't even considered yet. Immersive technology hardware will become something that all consumers use in their everyday lives.

In the healthcare field, we'll start to look at more

immersive technologies as another form of therapeutic treatment—a lot like how we think about medications today or therapeutic practices for rehab. Development of immersive technology in healthcare will be treated in the same way we treat new drugs. We'll have a lot of regulations and ethics in place to make sure that these solutions are being evaluated in a way that is safe for the end consumer.

In terms of aging and caregiving right now, globally a lot of countries, including the US, are in an aging care crisis. I think we're going to see a shift in the US where value-based care becomes the norm. We're going to see insurance companies covering a lot of immersive technology solutions because we will have proven that they save money and do the job better than current solutions. Aging care will be hugely impacted by augmented and virtual reality across all aspects of training and education, wellness, entertainment (probably less so in entertainment than other areas), and definitely therapeutics. From a public health perspective, augmented and virtual reality will be huge tools in population health, as they will help people stay healthier longer and out of the hospital.

What was the hardest obstacle you had to overcome?

The hardest obstacle I had to overcome was my own perception of myself as a CEO and getting to the place where I know I can be a CEO, I can be a founder, and I can grow a successful company. If I want a lifestyle business, I can have that. If I want a bootstrap company, I can have that. But what I really want is to fundraise and grow a company that takes on some initial risks to build for future technologies, and to fulfill that vision, I have to first believe that I'm capable of that. It's been really hard to be a first-time entrepreneur. I'm thirty, so I'm not that young, but I'm on the younger side for CEOs. I have to try to imagine that [being a successful young female CEO] is possible despite not seeing it in the world as the norm. It's really hard. I must continuously work to have confidence in myself, even if it doesn't always feel right, and to believe that I can do big things.

What was your big breakthrough?

My big breakthrough was seeing that this idea we had was actually a product, and a product with a market of buying customers. We almost went out of business. We were in the beta or premarket stage and didn't have a lot of paying customers, and we ran out of money. We had two choices: we could quit, or we could put what we had made on the market and see if it created enough traction with our customers to move forward. So, in January 2018, we took everything we had built and said, "We are out of beta and ready for market." We got enough traction to now have a healthy annual recurring revenue from our base of subscribers, and that's given us what we needed to say, "Yes, this is a business, and now we just need to figure out how to continue growing."

Why should younger women consider STEM and/or entrepreneurship as a career option?

Entrepreneurs build businesses that are reflective of their personalities, their interests, and their creativity, and we need people from all different walks of life and genders and races to bring their creativity and experiences into entrepreneurship. Right now, we still have set expectations of who become entrepreneurs, so we're not accessing all the voices and skills and experiences of people who fall outside of the standard profile of a white male founder. Women should consider STEM and entrepreneurship because a background in science and technical fields combined with creativity and problem-solving is a recipe for innovation. [By pursuing STEM and entrepreneurship], women will think of new ideas that can become businesses. They can grow those business in ways that provide value, and I hope they can become successful and earn enough money that they can return that investment back into other entrepreneurs whom they see value in.

Breanna Faye / rLoop

Breanna Faye is a trained architect turned experience designer, strategist, and technologist with a foundation in design thinking from MIT's Media Lab and global digital design agency IDEO. She is a cocreator of rLoop, an engineering organization utilizing blockchain technology to change the way remote-working or globally distributed companies operate. rLoop, which grew from a few members on a subreddit to twelve hundred engineers distributed across fifty-nine countries, is a crowdsourced organization using blockchain to build advanced engineering and design solutions, with an initial focus on the Hyperloop. rLoop teams won the Innovation Award at the SpaceX Hyperloop competition in January 2017 and the Best Design and Presentation Award from Dubai Future Foundation and ASITE.

Ms. Faye has worked on cutting-edge projects with a specialty in integrated mobility, smart homes and cities, Internet of Things, and future-foresight strategic deployment. She focuses on user experience through a design-oriented lens at intersections of systems, cities, and technology. Her projects range from urban scale to product scale, and include shared autonomous vehicles, the future of workplaces, blockchain technology, and GPS-enabled bike helmets. She has experience developing design solutions across North and Latin America, Southeast Asia, West Africa, the Middle East, and Europe.

What is your background?

I grew up in the rural Midwest, in Ohio farm country. I come from a blue-collar, conservative family. I started working at fifteen, basically as soon as I was legally allowed to hold a job. As a kid, I didn't really know what it meant to be an entrepreneur, but I was one. I did all the "kid" entrepreneur things: running a lemonade stand, mowing lawns, babysitting, and doing any kind of odd job I could get. That was my thing. Entrepreneurship has always been in my personality.

I had a goal to go to college. For many people, going to college is expected, but that wasn't the case for me. I was a first-generation college student. I worked hard in school and started supporting myself at a young age because I was always thinking about earning money and obtaining scholarships.

I attended the University of Florida for undergrad and MIT for graduate school, studying architecture and design at both. Before undergrad, I initially thought about going into business, but I ended up going with architecture because somebody recommended it as a great field for people who are both left-brained and right-brained—creative, mathematically minded people. That was me. I love the bigger-picture systems design that architecture speaks to, much bigger than a little building—more abstract and conceptual. I love the way you can use design to think about the world and any kind of system, whether it's building a human body, a city, or a product.

I was fortunate enough to earn full scholarships for both my undergrad and graduate degrees, and that afforded me a privileged lifestyle once I got out of college. I never had people supporting me, but I didn't have $200,000 in student loans that I had to start paying off right away. Being debt-free gave me a leg up to be able to pursue entrepreneurial projects and become involved with several startups.

Why did you start your company?

rLoop is a decentralized organization that started as a group of volunteers in response to the SpaceX Hyperloop competition. We've been working on four different products that are all within the built environment and are hardware-engineering-specific. We're most known for our work developing tech for the Hyperloop. rLoop is predominantly an organization of engineers, and I lead all things [dealing] with architectural design, urban design, or product design.

rLoop is a decentralized, crowdsourced community of engineers, architects, and designers that work on complex projects, such as developing advanced tech for the Hyperloop, a single-occupancy, human-flight device, and other passion-driven engineering projects for the built environment. How do you actually facilitate that [kind of collaboration] when you have a community of over a thousand remote contributors?

We currently utilize an ensemble of digital tools to work on projects—to ideate, to design, and to build things digitally and physically with team members who could be located anywhere in the world, as our community currently spans over fifty-nine different countries. I've been leading the design and strategy behind a platform that facilitates this collaborative effort, a smart-contract system that enables immediate remuneration of work done by skilled workers (engineers, architects, programmers, project managers, and so on). A smart contract is basically a way to make transactions—I give you this; you give me that in return. This person has worked this number of hours, and in return, they get x for it, which is a token instead of fiat currency. And instead of being facilitated by an outside entity or middleman, a smart contract is powered by code.

I got involved with rLoop because of an architectural design competition, sponsored by the government of Dubai, to design what Hyperloop stations might look like. The Hyperloop is the most advanced magnetic levitation technology that has been publicly administered. Because the Hyperloop is a totally new mode of transportation, when you think about the design, you get to think from the ground up. You can throw out all your past knowledge of how subways and airports and all our current transportation systems are designed. Instead of retrofitting the Hyperloop to fit current models, you can design an entirely new system for a new type of transportation. How to do that is basically what the competition was about.

I randomly got matched with rLoop, and we decided to participate together. They had the Hyperloop background, and I had the architectural design background. I led a small team in designing three different hypothetical stations. Everything was done digitally, remotely. We didn't know each other in real life; we were just strangers on the internet with different skillsets who were interested in doing this kind of project. So part of the challenge was finding a way for all of us to collaborate. The project required an immense amount of dedication and effort from a lot of different people—an aerospace engineer, me as a trained architect and product designer, and even people in PR and marketing.

After submitting, we were called back later that week and asked, "When can you get on a flight to Dubai?" because we were finalists. Amazingly, we ended up winning the Best Design award in the entire international competition, beating out well-known architecture firms that had dedicated internal teams working on the competition. And we were just a handful of probably fifteen people who had never met in real life, collaborating on this project remotely. They flew two of us to Dubai for the award, and that was the first time I met one of my teammates.

Now, I'm a core team lead at rLoop. I'm on the team leadership, but I'm not a cofounder per se. It's not that I'm *not* on the founding team; it's just a little ambiguous because we operate as a decentralized organization. Everybody's at the same ground level, and there are team leads based on their expertise. They provide guidance on anything within that topic. I'm lead of experience design, which is about the hybridization between digital and physical design.

Where does you interest in building companies and products come from?

I've always had a dichotomy of interests that have kept me torn between art and the more technical aspects of things. Creative and technical are both innate in me. As a kid, I loved art, and I loved math. I was interested in business, too. I even made—probably poorly—my own expense and earnings reports. I was counting pennies and dimes, but it was important to me to manage and understand the financial aspects of business. I say I'm a designer, but I've always worn a lot of different hats and had a lot of different titles.

Technically, I'm on the fringe of STEM. I work with people who are clearly STEM. They studied heavy mathematics or engineering, whereas I work in tech and am a very technical-minded person but am predominantly a creative. I never thought I would go into science or tech, but I think there's a beautiful synergy between creative people and individuals working in tech. Sometimes people who are technical have a hard time getting out of the minute-detail space, and the creative side teaches you to understand things at higher conceptual levels and think big picture. To have artists or designers who can come into the tech world and understand big concepts, like breaking down what the Hyperloop is and how the technology fundamentally works, is a beautiful thing.

How did you get your funding at each stage?

With rLoop, we're in the midst of this right now. We are currently in our private-sale stage launching our public sale, doing an initial token offering (ITO) as opposed to an IPO. So, partnering with organizations and experts to ensure SEC compliance, we're getting our funding through the blockchain space. The interesting thing about an ITO is that it opens the door to literally anyone in the world to invest, not just elite companies or venture capitalists (VCs). Anyone can be a VC in this day and age, with any amount of backing.

What advice would you give to other women in your field?

You have to put yourself out there constantly. Be proactive in learning how to position and grow yourself professionally. There's a lot you can do in terms of reading and studying to make putting yourself out there easier and more manageable. When I was younger and had professional questions, the internet was such a resource.

A lot of studies show that women tend to undersell themselves. You'll have a man and a woman with the same kind of experience and background, but the way the man presents himself makes him seem superior. I know I've been guilty of underselling myself because I don't want to say anything that's not factual. I just want to say, "This is what I can do. I'm confident in these things," in clear, black-and-white terms. But presentation is important.

You also have to learn not to be afraid, especially when you're in a room surrounded by men. And that is often how we find ourselves. I've been at conferences before where I was the only woman out of twenty speakers or where an entire eight-person panel was all men. As a woman, I think you have to position yourself and go after things a little bit harder. You need to get yourself in the door and then believe that you *deserve* that seat at the table, because you do. There are so many situations where you might question, *Do I really belong here?* and then you'll surprise yourself with how competent you really are among big names and people with twenty years of experience.

Always take a seat at the table, whether it's offered to you or not, because that's the only way you're really going to grow.

What advice would you give to STEM students?

STEM can seem like it's just hard sciences and hard math, which can be intimidating if you were never perfect at math. Don't let that deter you. Also, to get into STEM, you don't have to necessarily study STEM. I'm not advocating that you *don't* study STEM; just know that there are options, and you might end up getting to STEM via a different route. You can be a creative person and also get involved in STEM.

When I was younger, I was thinking of becoming an artist. I would have never in my life imagined that I would be working on such logically complex projects as I am today. I never would have ended up a chemist or a scientist, for example, because I never would have taken those courses, but I'm still in STEM, albeit a little on the fringe. There's no way I could compete with somebody who has studied engineering and wants to talk at the technical level, but on the conceptual level, we can have an amazing conversation and work on projects together, because we actually speak the same language.

What do you predict will happen in your field in the next five years?

I straddle a few different fields, but in terms of the built environment at large, what you'll start to see in the shorter timeline is more integrated tools and technology to better connect our cities and the way we move around, experience, and relate to them. There will be work on smart, connected transportation systems, making them more intelligent and giving them the ability to "talk" to each other, using Internet of Things (IoT) technologies, devices, and platforms. Blockchain has immense opportunities here. In a way, everything in our urban environment has been developed in isolation thus far—independent parts that we then have to figure out how to connect.

As far as mobility goes, there's something called end-to-end transportation. So, for example, right now, if I need to get from point A to point B, that might require three or four different [methods of transportation] to get there. I might have to walk to a train station, take the train, transfer to another train, and then walk or bike the rest of the journey. It's the same anytime you fly, right? You always need a car [or some other transportation] to get you to the airport and to complete the journey. In the next five years, what we're trying to do is to find ways to better tie all these loose ends and get people to the end of their journeys faster, in a more connected, cohesive manner.

Similar to the Internet of Things, the "transportation of things" will greatly improve the way people and goods move from one place to another. Ultimately, suburbs aren't sustainable. Things can't just keep growing wider and wider. Otherwise you'll have a five-hour commute.

End-to-end interoperability, data-driven analytics, and streamlined single and fleet management will improve conditions, eliminate wasted costs, and create a system of intelligent, interconnected transit. The advancement of Uber- and Lyft-type on-demand transportation services will become more widespread, regulated, and algorithmically effective, and they'll also be adopted by industries, private companies, and governmental organizations, not just consumers. This is a given and is already actively being pursued.

As transportation becomes more complex, diverse, and conglomerated (different modes of transit, more on-demand services/goods, more shipment of goods and people), there is an ever-growing need to address and reduce existing and future inefficiencies, both independently and within the entire ecosystem. Companies like Revmax, Uber, and Sidewalk Labs are working to carve a space that does exactly that—optimizing the placement and path of on-demand transit, projecting how certain mass transit shutdowns will affect given areas and commute times, minimizing wait times, reducing excess fuel burn off, and increasing ridership and reach to underserved communities in terms of mobility. Ten years ago in cities, you had to own a car or use taxis if public transportation didn't fit your daily commute. Now there are other options, and this connectivity and flexibility might expand out to smaller communities, like suburbs and rural towns, eventually.

There will be a huge improvement of AI integration and capability in self-driving cars down the line, when the cars actually start becoming intelligent decision-making machines. We will also start to see more of this [AI integration] adopted into our everyday objects and technologies, although it will be mostly invisible technology that drives these changes, meaning it won't feel like something out of a futuristic movie. Five years from now, we'll drive similar-looking cars, use similar-looking phones, but the technology will far surpass what we expect and see on the shelf today. Farther down the line, phones will start to look different as they become more integrated into wearables and smart textiles.

Farther out, there are some really interesting technologies when it comes to mobility—for example, moving cars on sleds (tracks) underground at fast speeds in order to circumvent traffic and ease the urban commute. Companies like Arrivo and the Boring Company are tackling this technology, which would address the problem of being stuck in traffic, especially if you need to get clear from one end of the city to the other, especially for a context like LA.

We'll also see new types of transportation entirely, such as Hyperloop transport or personal vertical takeoff and landing (VTOL) vehicles, which will require designers, architects, and urbanists to think from the ground up. That's a little farther out, probably like ten years, but you will definitely start to see these kinds of projects find footing. We haven't really done anything super innovative in terms of inventing something new since the airplane, so this new technology will be really interesting.

Is [the future] about moving cars? Is it about the elimination of cars? Or is it going to be about autonomous cars that are co-owned? You'll never need a car all the time, just for an hour or two of the day, so is there a way that you can co-own a car with three other people who have different transit times? In the Jovoto: Forework competition [in which we won first place for our Hyperloop station design], I explored exactly this situation, presenting a conceptual, human-centered approach and perspective of what shared autonomous vehicular transport might look like in ten-plus years and how it will affect our daily commutes, urban fabric/cities, and lives—having shorter work days, less time in traffic, less need for office space, etc.

In order for problems to be addressed and inadequacies to be improved, the overarching organization in charge needs to be aware. Technologies like embedded sensing, blockchain, and IoT can have a huge impact here—in the alerting process, logistics, delivery/transport—and almost all aspects of the supply chain can be connected and improved immensely.

What was the hardest obstacle you had to overcome?

It's never really about one obstacle, because there are going to be many. Lots of thought leaders and successful people talk about the importance of failure. The obstacles don't matter; it's about how you approach those obstacles and deal with them and what you do after. Sometimes you'll approach an obstacle and feel like you failed, and sometimes you'll feel like you succeeded. It's all about what you do after.

For me, getting to college was a big obstacle. There was no guidebook for somebody like me, with my background and where I came from. There was nobody to consult other than guidance counselors, because I didn't have family who had ever gone. So, figuring out how to get into college was a really hard hurdle. I was fortunate in a lot of different ways, but even things like "How do I live and eat?" were big obstacles in my life. The biggest obstacle for me, though, was getting over that initial barrier to learn how to get somewhere when I didn't know anybody who'd ever been there. "How do I do that? How do I get a job where I've never seen a woman working in that job before?" Right? Just take the small step.

Obstacles are always going to happen. I look for the path out. I work intensively and strategically. I was so focused on getting to college that things did work out. It's like if you apply for a hundred things, you're going to get a few of them. That's just how the odds are.

Working in the tech sector and jumping into technology startups, I have a different perspective of obstacles. You have such instability. You don't know where your next funding is going to come from. You really have to want to do [whatever you're doing] because you're giving up a lot—a stable life, job security, maybe benefits. You're also giving up having a clearly defined role: "I do this job, and every day I clock in, and I do this, and I know what that is." In the startup world and entrepreneurial world, you do everything. Sometimes you just have to be okay with a loose definition of your identity. I'm okay with wearing many hats, and even with knowing that not everything will work out, and I have to still want to do it enough to push through.

What was your big breakthrough?
That Dubai Hyperloop competition that my team won

was huge for me. At the time, I was actually working at a design agency in New York. I had a great cushy job, a leadership position. I could have stayed there in that much more clearly defined role. But we won this competition, and I felt validated as a designer and for thinking much more big picture. I'd always had these big-idea projects going on because I'm a creative and conceptual designer, and now all of that had been validated. So, winning that competition was a big turning point for me. It wasn't like I walked out the door the next day, but it created momentum.

Another super impactful part of that competition was that I worked with all these strangers whom I had never met who also believed in the project. I said, "Okay, look, there are other people just like you out there—people who believe that impossible-sounding things are possible and who want to see those projects happen." [The experience] reaffirmed things that I'd absorbed at MIT, which has this very entrepreneurial culture full of entrepreneurial spirit. I did that [kind of thing] for years, but you get sort of deflated with time. Projects don't work and so on. With this competition, I wasn't working on just a simple digital website startup; I had accomplished some really big technical projects. Winning reaffirmed and lit that spark [of entrepreneurial spirit] again, and when I felt the timing was right, I took the leap.

Why should younger women consider STEM and/or entrepreneurship as a career option?

I'm a trained architect, so it's technical but not technically STEM. However, I think there is a relationship between me and engineers and highly technical people who work with their hands and their minds. Some of the most artistic people I know are in STEM. There is such a beauty to understanding how to take an idea and actually make it. You can live in the space where you just have these cool ideas—it could be architecture; it could be with wood or with pottery or with electronics. But to then also have the technical skills to take that idea and bring it to life, to actually buy components and put them together and make something—whether it's a robot or a building—that is a really beautiful aha moment. If you stay just within the concept stage, you don't have the tools or the skills to actually build something. And if you live in the technical world, you're never trained to think big enough to come up with the ideas yourself.

When you have that fusion of mind and hand, it is like magic. It's not for everybody, but there is such a power in having the ability to both think and make, and that, I think, is what [the fusion of mind and hand] leads you to. Now, the tools are different, and the skills are a little different. I'm not a robotics person and couldn't do what they could do, and a robotics person couldn't necessarily design and build a small pavilion or building like I could.

But we're all builders and thinkers, and we can understand how you take these little, tiny things and build something bigger and kind of magical out of them. I think there's such a power and beauty in that.

If we talked about STEM that way—how some of these fields can give you the ability to create—it would resonate much more with kids, especially young girls, as opposed to just saying, "This is science and math." It's more, "This is how you animate things. You come up with an idea, you take your drawings, and you actually build them." That's super powerful, and I don't think that was ever conveyed enough to me in a school setting. Technology is just learning how to make things happen. It's basically magic.

When the artists and the electrical engineers and the architects and the chemists can all converse and understand each other at the conceptual level—that kind of teaching, that overarching means of understanding is profound. A good friend of mine, who's an astronaut and MD, and I were talking about the body and how it functions like a system. It's the same with buildings: you have your circulation, and you have all your components, or your organs. It's a system. I found it interesting that, though she and I are in very different fields, we speak the same language. I think that we need to focus more on speaking the same language. We need to talk more conceptually about STEM to kids and convey STEM fields in ways that are interesting and not just "This is a skill that you learn."

Marcie Terman / Panxora

Marcie Terman is a serial entrepreneur, market trader, and investor. She heads the communications division of Panxora (formerly First Global Credit), a multi-asset exchange that supports cryptocurrency trading. She is also a cofounder of Datafort, a company providing security solutions to London's financial service companies. She began her career in broadcasting at PBS's *MacNeil/Lehrer NewsHour*.

What is your background?

I was born in Brooklyn, New York, to an entrepreneurial family where my mother and father ran businesses together. They taught my sister and me to embrace productivity at an early age. There was a lot of joy in my upbringing, but that is not to say it was easy, because there was also a lot of really hard work. It was an unusual upbringing.

I graduated from NYU with a degree in film production, and after taking a gap year working as a deckhand in the Caribbean on sailboats, I went to work at WNET-TV as a news and documentary editor. I worked behind the camera in TV until I got the opportunity to work for an investment company in London that used proprietary AI to trade derivative contracts. I became the founder of a data security firm that financed the initial development needed to bring the Panxora Group successfully to market. Now we run a cryptocurrency exchange, a cryptocurrency-focused hedge fund, and an investment group that takes positions in early stage AI and blockchain companies.

Why did you start your company?

My business partner and I knew tech people who got into bitcoin early and were sitting on crypto-wallets crammed full of the stuff with no idea what to do with it. Because we actively traded markets, we offered to trade on their behalf and provide them with a way to profit from this asset that they had stockpiled. From there it was a short leap to consider doing this more formally and a huge leap in terms of the technology we needed to develop to deliver a dependable service to thousands of customers while managing the global risk of those customers and the whole business.

Where did your interest in building companies and products come from?

I've owned businesses of one kind or another since I was eight years old and started the best damn lemonade stand in Brooklyn. My mother encouraged me to be creative across the spectrum of whatever interested me, but she believed strongly in entrepreneurialism and self-reliance and passed that love on to me.

How did you get your funding at each stage?

We bootstrapped all the way. The original financing came from one of our existing companies. That worked the whole time the company was wearing her "training wheels." Now the prospects of the Panxora Group far exceed the fortunes of the company that helped found her.

What advice would you give to other women in your field?

Be true to yourself. Understand the areas where you excel and don't try to force yourself into a role that will not support your development as a professional. That said, you also need to be bold about stretching beyond your comfort zone, so you are always improving and striving to reach your potential. The midground between these two polar opposites will make for a rewarding career.

What advice would you give to STEM students?

I found a lot of benefit in a liberal arts education. I suspect this is not going to be a popular view, as liberal arts may not appear to directly apply to your chosen profession. However, having the broader understanding of the world that a liberal arts background provides will ultimately make for a more successful business life.

What do you predict will happen in your field in the next five years?

There will be extreme volatility and periods where the public view will toggle from hugely positive to extremely negative. However, acceptance is inevitable...resistance is futile. (Yes, that is a *Star Trek* reference.)

What was the hardest obstacle you had to overcome?

An early belief in striving for perfection. That caused me a lot of unnecessary pain and loss of momentum following the failure to achieve perfection. It took me a long time to learn that attempting to perform one's best instead of endlessly striving for an unobtainable ideal was the true path to growth.

What was your big breakthrough?

Embracing the truth that there are very few problems

that cannot be overcome, excepting death. This change in attitude allows you to achieve amazing results. Just amazing. Because instead of looking at challenges and saying to yourself, "I don't think this can be done," you immediately look at how you can do what you need to do. Much less wasted time.

Why should younger women consider STEM and/or entrepreneurship as a career option?

These are generalizations, and there will always be exceptions to these rules, but there are some distinct advantages that women have over men in business. For instance, it is my experience that women tend to analyze risk against potential outcome more realistically, which can help us make more-effective business decisions. We are not necessarily risk averse, but we demand that risk-taking offer significant upside potential before being willing to assume that risk and proceed down a particular path.

Fatema Hamdani /
Kraus Aerospace

Fatema Hamdani cofounded Kraus Aerospace in mid-2014. She is currently the company's president. Kraus provides unmanned aerial systems for defense and commercial use. Previously, she was vice president of mobile engagement and protection strategy at Syniverse, and partner at Ishi Systems. She was on Mobile Marketer's 2014 list of Mobile Women to Watch. She is an advisory board member of FinXTech.

What is your background?

I was born in India but grew up in Dubai. I always felt I was an expatriate there, so I moved to India to complete both a master's in hospital administration and an MBA. I didn't feel at home in India, either—only when I moved to New York nineteen years ago did I feel I had arrived home.

As part of my MBA program, I spent the last year helping

an IT consulting firm from Jersey City, NJ, set up their offshore center in India. That was a huge learning experience. I headed up the project, so I managed every aspect, from hiring to vendor selection, office setup, training, deployment, and making something operational from scratch. The firm then hired me into their New York office. Within a year of joining I made partner. I helped rebuild after the dot-com bubble burst and 9/11 hit some of their largest customers.

I subsequently joined VeriSign to help build out their enterprise offering, which was acquired by Syniverse. I continued to build the enterprise business for the world's largest mobile solution provider. I played an intrapreneurial leadership role and developed various ideas until I finally cofounded my own startup, Kraus Aerospace, in 2014.

Why did you start your company?

One of my biggest assets is the ability to solve business problems by applying technology. I am able to connect dots rapidly and have always focused on the "so what" when it comes to a cutting-edge technology and identifying patterns that exist across industries that technology can address.

I am motivated to create an employee-centric, living

organization that is known not only for the growth we achieve, but also for its mission of building an organization that keeps the good of the rest of the world in mind. Part of its mission is a working environment that creates joy, connection, and growth for everyone involved in building it. And it should have the ability to self-sustain and grow beyond the lifetimes of the original founders.

Where did your interest in building companies and products come from?

I want to leave a legacy and want to leave the world a better place. That has been a driving force in my desire to create companies and solutions that help solve for the big, hairy, audacious problems that abound around us. Once I recognized my unique gift of truly understanding technology and applying it to solve business problems and being able to do so faster than most others, I found real joy in creating something meaningful and solving problems. I like to create "ah-ha!" moments and change the course of how we operate.

How did you get your funding at each stage?

My business partner and I invested our own money, blood, and sweat to bring Kraus Aerospace this far. We now have an amazingly motivated and passionate team, a flying robot with ultra-high endurance capabilities in production, and a sales pipeline that is global and boasts of some of the largest customers that exist. We are now raising capital to stay ahead of the curve, maintain our competitive advantage, and expand the organization.

Knowing what type of partner you want to go to bed with is extremely important when it comes to picking funding alternatives, as the business will be living with that decision for a long time. It has the ability to change the course of a business both in a good and a bad way.

I have been selecting investment options based on the following four criteria:

- A partner who has the tenacity and vison to work with companies/products that require years of R&D and continuous commitment to R&D—who are not looking for the quick buck. They understand and support technology solutions addressing complex business problems of the future, not just an immediate, short-term win.
- Their other portfolio companies address complex problems needing multiple smart solutions to come together. Collaboration is key, as that is the only way to solve problems rapidly.
- The investor's commitment and ability to open doors that will help grow the business.
- The investor's commitment to supporting a company

that is an employee-centric, living organization that wants to build teams that are passionate, excited about what they do every day, and generate joy for themselves and others around them.

What advice would you give to other women in your field?

When one sets out to do something new and different and to create, one needs to first address their own fear. Sometimes that is the biggest limiting factor that will keep women, or any of us, from doing things we want to do, especially in the field of technology. Being okay with discomfort and the unknown when you are creating something new is a huge contributing factor that will lead to success. Know that breakdowns lead to breakthroughs, and hence, commit to allowing breakdowns to happen to truly create breakthroughs.

Also, always, always focus on the "so what" of technology. Don't build something just because it sounds cool.

Surround yourselves not just with promoters but with folks who will challenge you and force you to get out of your comfort zone. Talk to others who have a different opinion than yours so that you can challenge and better your solution. Don't ever stop learning.

What advice would you give to STEM students?

Science and engineering are vast fields that have various forms and implementation. Identify what calls to you, but also be aware that multiple facets of science and technology might be needed to solve a real-world problem, so having a baseline understanding of fields that might not be of immediate interest is important. Math and logic truly come in handy, as they will help form the basis of a lot of solutions. Know it's not just the traditional way of learning that can teach these concepts—for example, music is a good pathway to learning math and logic. Ask a lot of questions, especially the "why" behind things. Until you challenge the status quo, it's hard to know what needs to be learned in order to create something new.

Also, just because you didn't start by learning something in a formal manner from the beginning does not mean you cannot pick it up when you decide that you would like to pursue it.

What do you predict will happen in your field in the next five years?

I am in the field of aerial-based autonomous vehicles, and we aspire to build machines that will have nonstop airborne capabilities, utilizing multiple energy sources and the ability to navigate at various altitudes. The mission of the company is to "save lives" by shrinking the gap between data and decision, whether that be the life of a war-fighter, a civilian in conflict (Nigeria or Central African Republic or Syria), wildlife threatened by poachers, people involved in search-and-rescue or natural disasters, and so on. Our vision is about not just flying these robots but also integrating various sensors and other payloads that allow for gathering of data, then making sense of this data both in real time and via AI and machine learning to create actionable intelligence.

Finally, there is a huge gap that exists between space (satellites) and air (unmanned aerial systems), which is in the stratospheric layer. Our vision is to build a mesh network of unmanned aerial vehicles (UAVs) with non-stop airborne endurance. They will have a whole slew of capabilities, ranging from wireless services, intelligence surveillance, and reconnaissance to connectivity. They will be easy for operators to launch, maneuver, and maintain, and to swap out payloads to offer different types of services.

The satellite industry as we know it today needs to be challenged. It takes anywhere from nine to twelve years to launch a satellite, including cube satellites, and that is highly inefficient, seeing how fast the technologies these satellites need to support are evolving. For example, if our mesh network of UAVs in the stratospheric layer had been operational, we would have had Puerto Rico up and running on wireless services within weeks of the Hurricane Maria disaster.

What was the hardest obstacle you had to overcome?

Being okay with the unknown and uncertainty. From a very young age, we are all taught a certain way about how success is measured or timelines are imposed. These come from societal norms that we grow up with. It is hard to overcome the fear of not conforming to them and somehow being left behind. What if you are not married by a certain age, or have not had kids, or a house with white picket fence? This is especially true for women, due to the ever-impending biological clock. I had to deal with this and break through it.

Some of my hardest obstacles were my own self-limiting beliefs: what if I am not good enough, what if I fail (it is actually okay to fail, as our failures are our biggest teachers), what if I am not able to keep up the personal commitments that I have toward my family, and so on. They are just that, self-limiting beliefs. Our ability to go through our fears and come out at the other end is what is key.

What was your big breakthrough?

Not coming from scarcity and truly creating from a place

of abundance. This is extremely, extremely important. Every time we have hit an obstacle, we have had success when we have created from abundance versus taking the answer "It can't be done." Whenever we have been told, "It can't be done," we pushed ourselves to come up with solutions. When large aircraft manufacturers around the world told us what we were thinking was not possible, we went out and proved them wrong. Now white papers are written about our technology.

A second breakthrough: hedging has diminishing results. One of my biggest breakthroughs was when I left my full-time, extremely well-paid job and very settled life to pursue my passion and dedicate my entire time and attention to my startup. It was extremely scary, but one of the best decisions I ever made.

Why should younger women consider STEM and/or entrepreneurship as a career option?

Some of the biggest and most successful leaders attribute a lot of their successful decisions to a combination of prior experience culminating in knowledge, which further enhances their inherent ability to make decisions "instinctively." This is one of the attributes that leads women to be extremely successful in any career that they choose, but more so in entrepreneurial paths.

One of the first people to launch a commercial satellite was a woman, and several similar examples of women pioneers in tech come to mind. The sky is not the limit when a woman applies herself, hones her instincts, and trains to help make better decisions. Additionally, our natural nurturing instincts lend themselves to growing and nurturing powerful teams that are the cornerstone of any successful organization. It is not just women, but diverse teams, that allow for all of us to come together and build great things.

Cristina Dolan / iXledger

Cristina Dolan is cofounder and COO of iXledger, an alternative insurance marketplace based on blockchain. Previously, she cofounded internet service provider OneMain (later acquired by Earthlink) and served as an executive at Disney, Hearst, IBM, and Oracle. She is a member of the Forbes Technology Council and has served as both vice chair and chair of the MIT Enterprise Forum in New York City. An engineer and computer scientist, she has a master's degree from the MIT Media Lab. In 2013, she initiated an award-winning global competition for students, Dream it. Code it. Win it. Her TEDx Talk, "Just Solve It!," has been viewed close to a million times on YouTube.

What is your background?

Multiple facets to my background have helped me become a successful entrepreneur, solutionist, and futurist. In college and graduate school, I studied electrical engineering and computer science, which enabled me to understand everything from power to circuit design to communications protocols. At the MIT Media Lab, my studies were focused on interactive storytelling, which took into consideration everything from the technical aspects to the narrative. My career has spanned a variety of industries from telecommunications to media to financial technologies. I've worked for IBM, Oracle, Hearst, and Disney, serving in a variety of roles, from head of technology and development to lead sales and business development roles. This background enabled me to join founding teams to launch new products and companies, and my engineering background trained me to think like a problem solver, which helped me identify and commercialize new technologies throughout my career.

Why did you start your company?

I started OneMain because I saw an opportunity to address a need in the market. Working with a solid team has been critical [to our success]. Entrepreneurialism in tech is much easier today because there is a standard stack of technologies that can be utilized to build industry solutions. Years ago, companies would hesitate to use technologies developed by startups or small companies, because it would represent risk. The interesting part of the puzzle is understanding customer needs and building solutions that address those problems.

Where did your interest in building companies and products come from?

As an MIT alumna, it is hard not to have an interest in building products and companies; it is part of the DNA of the school. I was fortunate to have been there just as the web browser was evolving, which fueled a tremendous wave of innovation that in turn fueled my career. I went on to head up technology at Hearst and ABC/Disney, building some of the first consumer websites.

The MIT Media Lab was an incredible place to study, and I was inspired by the passionate work of researchers and students tackling some of the most complex and challenging problems. Many of my peers have built amazing products and companies as well. I also credit my involvement in blockchain to a visit to the MIT Media Lab six years ago, when I was fortunate to hear a presentation around the ability to immutably store a few bytes of information as a time-stamped proof.

I did also come from a family of very successful entrepreneurs. My father was a chemical engineer and a successful businessman who encouraged me to pursue engineering.

How did you get your funding at each stage?

Throughout my career, I have worked with friends and

family, angel investors, family offices, venture capitalists, and the public markets [to secure funding].

OneMain was initially funded by an investor, and we later went out to the public markets and completed one of the most successful IPOs of the time.

With Dream it. Code it. Win it.—the successful not-for-profit I launched in 2013 that awarded students for solving interesting problems with code—the initial funding came from MIT alumni who believed in the idea of competition to encourage a diverse group of students to pursue engineering studies. After the successful first event, it was much easier to raise money from sponsors using the metrics from prior events.

What advice would you give to other women in your field?

Get experience and take advantage of online resources to continuously learn and build your network. If you want a seat at the table, you need to come with a level of knowledge and experience that will give your customers, investors, and employees the confidence to believe in your abilities.

What advice would you give to STEM students?

Science and engineering are difficult subjects, and many

female students quit or give up. Unfortunately, the statistics aren't improving. The training you receive as an engineer and the analytical skills you sharpen through engineering programs are applicable in any industry. We live in a world where technology is part of everything we do, and if you don't have the skills, you don't get to be at the center of building the future—that is where you will find the most exciting and rewarding career opportunities.

What do you predict will happen in your field in the next five years?

We are already seeing innovation happening at hyperspeed. Blockchain is evolving as a way to address compliance and regulatory requirements, and it is creating new, efficient business processes that will be as disruptive as the internet was to media and as e-commerce was to retail. The technology will continue to evolve with development tools and data analytics. If you think about the early days of the internet where the first consumer websites were focused on driving as much traffic as possible to the home page by buying traffic from search engines, the models didn't work or scale. The next generation was about the customer journey with intent, and companies like Google and Facebook have taken over close to 80 percent of the market by capturing these journeys and offering the "intent" to buyers on a marketplace.

Blockchain also captures journeys—the journey of a coin, the journey of a customer, or the journey of a claim. This slice of data currently doesn't live in a single system. Today, this data is spread out through siloed systems that belong to other organizations, and reconstructing the data from these systems isn't practical. The ability to capture and view this data will create new opportunities and efficiencies.

What was the hardest obstacle you had to overcome?

I try not to think about obstacles as "hard," because it tells your brain that it is okay if you don't accomplish your goals. I like to focus on where I am going. At times it has been a bit like the boiling frog, where I didn't realize how hot the water would get until it was boiling. Once you are past the point of pain, it is always best to focus on the goal you accomplished.

For instance, public speaking is a requirement for any leadership role, but it creates a lot of fear for many people. While early in my career I wasn't as confident, I made it a point to take the opportunity to speak on topics I understood well. Now my prior successes have given me the confidence to overcome my fears.

What was your big breakthrough?

When I created Dream it. Code it. Win it., I decided that solving an interesting problem with code would be the best way to judge student submissions and create an event to showcase the winning entries. At that time, I had no idea that problem-solving is hard and is a learned skill that few people are fortunate to acquire. Understanding the importance of problem-solving and its link to innovation has been critical to my career.

Why should younger women consider STEM and/or entrepreneurship as a career option?

We live in a world powered by technology, which is changing everything we do. Those who are involved with technology and engineering will get to be a part of this revolution. There are many opportunities being created at the center of this wave of innovation. There are no shortcuts, though. Women won't have the career opportunities, management positions, or appointments on boards without the skills.

Melonee Wise / Fetch Robotics

Melonee Wise is the CEO of Fetch Robotics. She began her career at Willow Garage, where she led a team of engineers developing next-generation robot hardware and software. In 2015, she was one of the *MIT Technology Review*'s "35 Innovators Under the Age of 35." She has also been named on the *Silicon Valley Business Journal*'s "Women of Influence" and "40 Under 40" lists. In 2017, *Business Insider* named her as one of eight CEOs changing the way we work. Under her leadership, Fetch Robotics won the MODEX Innovation Award for the materials handling industry, and was named to "RBR 50," a listing of the fifty most innovative robotics companies in the world.

What is your background?

I attended the University of Illinois, where I earned a bachelor's in physics engineering and mechanical engineering and a master's in mechanical engineering. While at university, I interned at Daimler-Chrysler and Alcoa and worked at Honeywell. I started the PhD program in mechanical engineering, but about six months after passing the qualifying exams, I received a compelling offer from Willow Garage. In the span of a week, I dropped out of the program and moved to California to begin working at Willow Garage, which was a really small company at the time, with only three or four employees.

After Willow Garage, I started Unbounded Robotics, a Willow Garage spin-off, with three coworkers [from Willow Garage]. Due to some complications around the spin-off agreement, the company shut down. Around that time, the startup Fetch Robotics was looking for roboticists, and the four of us from Unbounded Robotics were all hired. Tech-RX, the venture capital firm that had started Fetch Robotics, provided the groundwork, and then they stepped back and let us roboticists take over and start running the company.

Why did you start your company?

Willow Garage was an exciting and interesting place, but a lot of our projects never saw the light of day. I got tired of building things that no one ever used. I wanted to build something that would be useful in the real world—I wanted to have an impact. It was really hard to have the kind of impact I wanted at Willow, so I started Unbounded Robotics.

Where does your interest in building companies and products come from?

I want to build things that are real. At Willow Garage, the challenge was decision paralysis. I wanted to make a decision and go after it, and I had cofounders who wanted to do the same thing.

How did you get your funding at each stage?

At Fetch Robotics, we raised our seed funding in October or November 2014. Through Unbounded Robotics, I had connections to the venture community that allowed me to meet people interested in our ideas. I pitched to quite a few people, and eventually we got a term sheet and then seed funding from O'Reilly AlphaTech and Shasta Ventures.

Series A was a little bit different. I had become acquainted with the CTO of SoftBank, and he was interested in what we were doing. He sent someone to talk to me. I went back and forth with various people from SoftBank in a series of meetings. At one point, it seemed the negotiations would end without us reaching an agreement, but eventually, we came to terms. Their investment in Fetch Robotics was a balance sheet investment, as their investment arm, SoftBank Vision Fund, did not yet exist.

Two years later, in summer 2018, my COO and I went

through the traditional Series B venture fundraising process. In fundraising, you have to be strategic. You must try to create urgency and bring a variety of communities together in a very short amount of time. You have a lot of meetings and must drive people to make a decision. We had quite a few people who were interested [in investing], so it was about choosing the right partner, which we decided was Sway Ventures.

What advice would you give to other women in your field?

Everything takes longer than you think it will, and everything is harder and lonelier than you would ever believe. Becoming a startup CEO is a hard path, and it's not for everyone. If you're doing it for glamour, success, and media praise, you're in the wrong field. You have to be ready to make hard decisions and take ownership of those choices. A lot of people struggle with taking ownership of the failures along with the successes, but it is necessary to be successful.

What advice would you give to STEM students?

If you're from the Midwest, come to California, come to Silicon Valley, so that you can experience the engineering culture. This is especially important if you're a woman. Gender discrimination does still exist in the Valley, but compared to the Midwest, the Valley is the most open and accepting place for women engineers.

Get internships. Get a job at a big corporation. I worked at big corporations in graduate school, and I learned that they were not for me. Explore different types of jobs. Try on every hat you can.

What do you predict will happen in your field in the next five years?

From an outsider's perspective, not much will happen. Some academics say that the problems of outdoor navigation, localization, and mapping are solved, but none of those fields are solved. These are difficult problems, and the needle won't move substantially in five years. Five years simply isn't a long-enough measuring stick. You might say it takes five years to become a success and ten years to become an overnight success. We still might not have fully autonomous cars for another fifteen or twenty years. Twenty, thirty years is the time frame you have to look at.

What was the hardest obstacle you had to overcome?

The hardest obstacle was learning everything I didn't know and admitting that I didn't know certain things.

What was your big breakthrough?

My big breakthrough was learning that I was passionate

about challenge. I needed challenge. In my career, I rein-vented myself every five years. First, I was a mechanical engineer, then I was a programmer. As a CEO, I find that every six months the job is totally different. I'm solving new problems all the time.

Another big breakthrough was learning that I have to own my decisions. People tell themselves stories: they're a victim; there's a villain. None of these stories make sense to me. I work to accept that things are what they are—that can be a hard place to get to. Over the last six years, I've become very introspective and try to be as self-aware as possible. It makes me better at my job.

Why should younger women consider STEM and/or entrepreneurship as a career option?

People need to find what they enjoy, but culturally, we don't give women and certain other groups the oppor-tunity to discover whether they like entrepreneurship or STEM. I read recently that only 7 percent of parents would encourage their daughters to become engineers.*

Society has built a lot of walls. We channel people into certain activities and fields instead of allowing them to

---

* Julia Glum, "Girls In STEM: Parent Stereotypes May Discourage Daughters From Science, Technology, Engineering, Math Careers: Study," *International Business Times*, April 24, 2015, https://www.ibtimes.com/girls-stem-parent-stereotypes-may-discourage-daughters-science-technology-engineering-1895719.

explore. We've built up strong archetypes around what it means to be an engineer: exceedingly smart, completely antisocial, low EQ. This stereotype has driven people away from the field. How we market engineering has to change. We have difficult problems, and we don't have enough engineers in the US or in the world to solve them.

Is there anything else you'd like people to know?

I'm an engineer and an entrepreneur, not a woman engineer or a woman entrepreneur. Being a woman has nothing to do with my skill set, but it seems that the two are eternally linked.

Elizabeth Rossiello / BitPesa

Elizabeth Rossiello is the CEO and founder of BitPesa, a digital foreign exchange and payment platform that leverages blockchain settlement to lower the cost and increase the speed of business payments to and from frontier markets. Founded in 2013, BitPesa was the first company to leverage digital currency to significantly lower the cost of making remittance and business payments to and from sub-Saharan Africa, as well as the first blockchain company to be licensed by the UK's Financial Conduct Authority. BitPesa is now a market-maker in every major African currency and facilitates payments into G20 currencies directly. Ms. Rossiello expanded the company from Nairobi, Kenya, to operations in eight markets across Africa and Europe. Before founding BitPesa, she was a rating analyst for microfinance institutions across sub-Saharan Africa, consulting for the Grameen Foundation, the Bill & Melinda Gates Foundation, and the Acumen Fund, as well as working with regulators and policymakers on legislation for financial innovations. Elizabeth sits on the Investment Committee for Bamboo Capital and the World Economic Forum's Future Council on Blockchain. She holds an MA in international business and finance from Columbia University.

What is your background?

I come from a traditional finance background, and never really thought that this would be my journey. I also believe when you face a compelling problem, it is your responsibility to solve it. When I first moved to Kenya, there were so many big, hairy problems that I knew would be solved using technology that I had to become part of it. Even though I don't have a particularly strong technology background, you learn how to hire and collaborate with people with complementary skill sets. For example, our COO, Charlene Chen, has both a personality and skill set that is so vastly different from mine—that's why we work so well together.

Why did you start your company?

After working as a rating analyst with some of the biggest microfinance institutions across sub-Saharan Africa, as well as with investors in the space such as the Grameen Foundation, the Bill & Melinda Gates Foundation, and the Acumen Fund, I saw the difficulties small financial institutions faced finding liquidity in local African currencies. I repeatedly saw funds and institutions have to exit the market because of foreign exchange FX loss or lack of local currency infrastructure. It seemed like such an obvious problem that someone had to be solving it. I was also working with regulators and policymakers on legislation for financial innovations, but that wasn't enough for me—I knew that I had to do something out of the box, and urgently. In 2013, cryptocurrency had just started taking off, and I knew that it was the right tech solution for this problem.

Where did your interest in building companies and products come from?

I didn't really think of myself as an entrepreneur in the beginning—I had always worked in corporations and finance before. However, after my experience working in Kenya, I was presented with such a compelling problem that I knew I personally had to do something. That's how BitPesa started.

Tell me about your funding challenges.

Raising money is never easy, but we all have to do it at some point. At BitPesa, we are lucky that we are in a hot area—blockchain—and we're presenting a scalable and true-use case for cryptocurrency. To this end, there were a number of top investors in the space interested in us, such as Pantera, Bitfury, and Digital Blockchain Group. Having these strategic investors proved invaluable later, and they are still great allies for us. However, we were not lucky in that I was a female founder based in Africa, and that made the fundraising process that much harder. Women only receive 2 percent of venture funding in the

US, and that problem was only amplified by the fact we are a frontier markets company. We pushed through and to date have raised more than $10 million in funding.

What advice would you give to other women in your field?

The blockchain industry is a combination of two very traditionally male industries—finance and tech. But it is an entirely new industry, and this is an opportunity for women to break ground, and they already are. The decentralized communities that exist across the space make it easy to reach out and connect with women outside of your hometown or your own office. This decentralized support has been pivotal to pooling resources, sharing knowledge, and helping women climb up faster to a more equal place. Many men in the industry still think that speaking about gender is unnecessary and that, if a woman is qualified, she will simply rise to the top as a man would without the need for any additional support. These men do not understand the implication of systemic privilege and bias that hold even the most qualified women back. We still need to speak out and proactively pull women up to the top.

What do you predict will happen in your field in the next five years?

Currently most blockchain applications are being developed within silos. Blockchain has been such a fad that people are depending on it as the single technology that will usher in a new age, ignoring the fact that there are a number of other rapidly developing, emerging technologies, such as AI/machine learning and quantum computing [that will have substantial implications]. For example, if quantum computing is commercially viable, then it will render all our current encryption methods completely useless.

In the next five to ten years, there will be more synergistic opportunities and codevelopment between blockchain and other emerging technologies of the Fourth Industrial Revolution.

Why should younger women consider STEM and/or entrepreneurship as a career option?

Why not? If there are glass ceilings and old boys' clubs in traditional industries, then entrepreneurship is an opportunity.

Olga Egorsheva / Lobster

Olga Egorsheva is the CEO and cofounder of Lobster, an AI-powered platform that enables brands, agencies, and the media to license visual content directly from social media users and cloud archives. Her first experience with entrepreneurship came in the form of a family photography business, which she founded with her father in her native Russia in 2005. From there, she studied and worked in a variety of cities, including Oslo, Paris, and Bonn. After receiving her MBA, she settled in London. Her speaking opportunities have included TechCrunch Disrupt, the IAB European Advertising Congress, Cannes Lions, Data Natives, Pirate Summit, and Social Media Week. She has degrees from ESCP Europe and BI Norwegian Business School.

What is your background?

I grew up near Moscow and got my bachelor's degree in economics at Moscow State University. While at university, I helped my father start his latest business venture, which was all about photography and enabling photographers to build their own customizable home studios by combining equipment, computers, and software in a smart way. That was my first immersion into entrepreneurship, and funnily enough, it is related and helpful to what we're doing now at Lobster.

Once I'd finished my degree and gained a bit of experience with my father's venture, I moved on to do a two-year MBA in international management and marketing at the BI Norwegian School of Management in Oslo and ESCP in Paris. When I was in my final module, I met a great professor in entrepreneurship, and I kept running after him, saying, "Hey, I want to work with you! Do you have any jobs?" He was one of the leaders, one of the tops in management, at DHL at that time. Eventually, close to the end of my MBA, he offered me a position. I stayed at DHL for about five years. As with many of us, I had this perception that even if you've always dreamt of starting your own company, you should first get professional experience in a bigger company with properly built processes and clients in marketing and finance before going out and starting something of your own. I now think that's actually a controversial thought. You can go and start something straight out of uni these days. You can go whatever route.

I started with DHL in Paris and then moved to the headquarters in Bonn. I was working in a sort of corporate entrepreneurship role. I looked at new markets in Africa,

Singapore, and Russia in terms of geography and services, built business plans with the finance teams, and launched new services or business units in those markets. In my second year I went to Russia to start a new business unit for DHL from scratch. The business unit specialized in providing a new service for e-commerce delivery that should be more economical than express courier delivery while also being reliable and having great customer service, great notifications, great IT, and transparent service terms. So, not your average postal delivery. My team built exactly that from scratch, and when it became profitable, I decided it was time to realize my dream to go and start something on my own. One of my cofounders experienced the same kind of moment in her life around this time, and we did it.

Why did you start your company?

Lobster actually came out of one of my pain points while working at DHL. Along with business planning for new products and new markets, I was also doing strategic marketing, and that included anything from decks to conferences. I was always struggling with the stock photography because it was all European- or American-centered—you know, blond family with two kids or two men in front of a whiteboard. I thought, *How am I supposed to work with Middle Eastern, Singaporean, and Russian markets with this stock imagery?* And that led me to think, *Why can't you use social media images and videos from real people in those countries?*

I realized that copyright issues prevented us from doing so. From my earlier work with my father's photography venture, I also knew of the pains photographers faced in monetizing their images and dealing with piracy. This all led me to the idea of providing a platform that allowed for the ability to source rights from individual authors and to pay them in a quick, safe, and legal manner. My cofounder Maria has a designer background and experience working with images, and the other tech cofounder, Andrey, is a keen social media photographer himself. It all kind of came together.

Where did your interest in building companies and products come from?

My most important inspiration was seeing my father start businesses as I grew up. I watched my father evolve from a scientist—educated, interested, and passionate about what he was doing but still working for others, for research institutions—into an entrepreneur, building his own companies around something new. Computers were just appearing in Russia after the Soviet Union collapsed in the '90s, and he took advantage of this opportunity to build computer businesses. He built successful businesses on his own, according to his own vision, and created new value in the market. Throughout my teen years [and at university] I did a lot of different things, from journalism to writing to maths, but by age eighteen, when it was time to choose my field of study and passion, I was pretty sure I wanted to be an entrepreneur.

How did you get your funding at each stage?

When we first started, we bootstrapped off our own savings, hiring one or two developers to help out. Once we'd built a prototype, we applied to a few accelerators, and we were invited to the Wayra UK accelerator. Wayra UK is one of the most well-known accelerators in the UK, and they're also active in Europe and Latin America. They belong to Telefónica, one of the largest mobile networks, so they specialize in everything that is mobile consumer-driven. They gave us our first investment—about £50,000, in a convertible note—and also quite a bit of structure. Pitching in their finals was one of the best decisions we made at the start as first-time entrepreneurs. Getting that cash meant we were able to hire some people and experiment with marketing. Our acceptance into the program also provided the added benefits of weekly traction meetings, investor speeches, in-house accounting and legal advisers, and so on.

After the demo day and TechCrunch Disrupt EU in London, where we were among the fourteen European startups to pitch onstage, we immediately raised a £100K angel round from a group of business angels. With that funding we started getting the first content and users on the marketplace, and Lobster started growing nicely.

Less than a year later, we raised £350,000 through crowdfunding and angels again. That money went to a large extent to continuing to grow as a community, but it also went toward our first client experiments and artificial intelligence development. We started working on artificial intelligence development because we realized that the platform will be much more valuable to clients if we, for example, tag the images and videos with AI, allowing clients to do things like filter according to quality or desired diversity elements.

Then, last year, 2017, we raised £1 million through small VCs and angels. We've used that money to further develop artificial intelligence tools and many other exciting things.

What advice would you give to other women in your field?

Feel equal in this space. Believe that you are on the same level as your male counterparts. Whatever your gender is, when starting a business, what drives you is your dream or vision, the business plan, the market opportunity. If you don't cultivate that bias [against women] in yourself, then, in my experience, others won't have that bias against you either. I probably have a slightly different story, though, because I was very much inspired by men, by my father, and I grew up in a country where, because of socialism, men and women were equally hardworking. There were always women doctors, teachers, finance directors, and company directors. With Lobster, I've never felt as if my female cofounder and I were different from our male cofounder, even in the accelerator. I think it helps a lot to not feel like you are somehow different or disadvantaged, because then others won't see you that way either.

What advice would you give to STEM students?

STEM, entrepreneurship, and everything in that intersection are some of the best things you can do right now. Artificial intelligence in particular is a very exciting field. Sometimes teenage students studying artificial intelligence come over to the office to see what we're doing, and I always encourage them to keep proceeding in this field because artificial intelligence is starting to become like a prerequisite, a natural ingredient in any business or private service that is being built.

Whatever you study, be prepared to keep learning and adapting. People ask me sometimes, "Will AI take our jobs?" In my opinion, technology doesn't take jobs, but it does *change* jobs. Instead of fearing the changes, I think it's most important for students and professionals to learn to adapt and work alongside technology. With lifelong learning, you can understand how your chosen field is changing and move your skills as a worker or your ideas as an entrepreneur to keep pace with the changes or even lead them.

What do you predict will happen in your field in the next five years?

In the creative and media field, I think we'll see developments in both artificial intelligence and machine-learning-based technologies that will allow for advancements in everything from image curation to creative scenarios in production to music composition to computer-generated or computer-adjusted images and videos.

I also think humans and technology will learn to understand each other better. For example, we're working on something at Lobster so that creatives don't have to use technical keywords and can instead say, as sometimes they do, "I want something that looks as fresh as a breath of air." And then the machine will understand them. Creatives who are tech-savvy and utilize all the available tools can then spend more time brainstorming and coming up with their next great idea to conquer our hearts.

I also see a lot of democratization and individual-economy empowerment through technology. Sometimes we say Lobster is Uber for stock photography, and more and more industries are moving toward similar models that empower and protect individual contributors or contractors—an individual driver or an individual homeowner or an individual social media user. Block-chain is helping with this shift because you don't need physical things like books of licenses, transactions, or invoices anymore. For every individual contracted, you can make all transactions public and accessible, which increases transparency and helps empower individual contributors.

What was the hardest obstacle you had to overcome?

Early stage funding is always an obstacle. It's super important, and it's super hard to obtain, especially for first-time entrepreneurs. You have to convince people to trust you, your idea, and your team early on. Even though we easily raised our first £100K after the demo day, it was a big obstacle. Then, especially if you're working on the business-to-business side of the market, another big obstacle is getting the quality mark, increasing awareness, and gaining the trust of your big partners or clients so they'll buy from you.

What was your big breakthrough?

There were a couple. One was the TechCrunch Disrupt Battlefield presentation early on. It triggered a lot of press, from TechCrunch itself to smaller media outlets, as well as investor interest, even from investors who didn't invest at that moment but did later. Because of the attention from that presentation, the giants like YouTube, Facebook, and Instagram got in touch and sat down to talk with us and try to understand our business and embrace it in some cases, allowing us to build partnerships and relationships.

The second big breakthrough was our £1 million funding round and our following presentation at the Cannes Lions, an annual festival for creative fields, advertising, and media held in Cannes. It is like South by Southwest or like the Oscars for advertisers with a huge creativity fest in addition to the Lions awards, which are the biggest awards in advertising. All the big agencies, creatives, and industry people attend, and thanks to funding from the New York-based R/GA Ventures, we were able to go and present our product for the first time in front of the crème de la crème of the creative industry from around the world. Europe, US, Asia—everyone was there.

If TechCrunch Disrupt was a breakthrough moment for us in terms of technology market recognition, investor recognition, tech press recognition, and subsequent investor conversations, then Cannes Lions was a breakthrough in terms of putting our high-tech, innovative product in front of the world's creative leaders and reaching their eyes and ears. When you're a startup and you're at the beginning of the journey, however cool and innovative you are, it's hard to get on the big brands' radar and get them to trust your brand and start working with you. Cannes Lions allowed us to start those relationships and overcome that obstacle.

Why should young women consider STEM and/or entrepreneurship as a career option?

Unless you have another passion for literature or music or something you want to do for life, STEM and entrepreneurship are some of the most prominent and developing fields. If you want to create something, help build the future, and always be part of the latest happenings in the world, you should consider STEM and entrepreneurship.

STEM in particular is becoming embedded in our daily lives. In the UK, at some schools, kids are being taught how to program at roughly the same age they're taught how to write. They're learning how to build little robots, how to program software and hardware. For the next generation, being able to understand tech and take a dream or idea and build it is becoming as natural as writing down your thoughts when you have them. You get the idea, you test it, and you create something, and creating something very often means programming something these days.

With the new generation, I also think this issue [of gender bias in STEM] is beginning to disappear. Boys and girls are not doing different things anymore. You can see it with those robot-building classes, where schools are doing a great job of teaching boys and girls all together and to the same extent. As a result, not only are young women encouraged to pursue STEM routes if they're interested in them, but young men are working side by side with women, so they don't have the bias that it's *their* field. So I think the new generation is more immersed in STEM and also more gender agnostic.

Leanne Kemp / Everledger

Leanne Kemp brought her extensive background in emerging technology, business, jewelry, and insurance to the founding of Everledger, a firm that drives transparency and trust along global supply chains using emerging technology. Ms. Kemp, the company's CEO, previously founded three Australian startups, including Absoft Queensland, Fastcards Pty Ltd., and the Great Australian Survey Company. Through these companies, she introduced and patented transformational technology ideas, including Fastcards (a smart card with managed identification services), and commercialized Multicard (credential verification for consumer, governments, and corporations), alongside various solutions to streamline inventory and supply-chain operations.

Her entrepreneurial success has led to her appointment as the Queensland Chief Entrepreneur for Australia in 2018, a role intended to further develop the startup ecosystem, attract investment, and support job creation throughout the state of Queensland. She is the first female entrepreneur to hold this position. Ms. Kemp is an appointed member of the World Economic Forum's Global Blockchain Business Council and a cochair for the World Trade Board's Sustainable Trade Action Group. She is also on the IBM Blockchain Platform Board of Advisors. Her awards include Innovator of the Year 2018 at the Women in IT Awards (London) and the Advance Global Australian Awards 2018 for Technology Innovation.

What is your background?

I've spent the better part of my career in software development, and I'm a serial entrepreneur. At university in Australia, I studied commerce, and I'm the founder and CEO of Everledger, which began in London in 2006.

Why did you start your company?

Two main threads came together at one time. In the mid '90s, I worked in radio frequency identification (RFID), which is track-and-trace technology across supply chains. That technology was expected to be a disruptive force within supply-chain traceability.

Then in 2008-2010, we started to see cryptocurrency and the effectiveness of blockchain becoming potentially disruptive. I saw I could marry my previous experiences together in disruptive technologies to be able to solve the very real problem in supply chains.

When we think about the types of applications that connect business networks together, there are procurement systems—applications like Ariba. But there is no application that is a "platform of provenance"—that answers the question, how do we track objects from the source of their origin through their lifecycle, to the point where that object either becomes consumed or ends up in a circular economy? A number of supply chains are opaque or conflict-driven, such as diamonds. [I asked], how do we build a platform of provenance to combine emerging technologies like blockchain, smart contracts, machine vision, and artificial intelligence to enable traceability and transparency, and to de-risk supply chains globally?

We began with diamonds. We could take a nondestructive set of methods to identify a diamond at a thumb-printing level of DNA, then be able to match the digital identity of that stone into the blockchain. We could match the physical stones and digitally record by means of scientific identity, and then enable the traceability of those stones across the network.

How did you get your funding at each stage?

The company's been generating revenue since its inception in 2015. We created a Series A investment round, led by Fidelity, with Bloomberg and Rakuten and a number of other strategic investors participating. We raised $10.5 million in February 2018.

What advice would you give to others in your field?

Enable a purpose with vision, and complement that through persistence and hard work.

What advice would you give to STEM students?

It's all about application. It's about how we can execute and apply technology in a way that serves industry. When operationalizing any set of knowledge, it is important to understand where the technology will play a role in solving challenges or problems within certain industries and fields. The technology cannot be technology for its own sake, for the sake of science or theory. If you are able to take your knowledge and apply it, extending that into the future-possible, then you are proving your own worth in terms of your background study or expertise.

What do you predict will happen in your field in the next five years?

We should be thinking about innovation from within the product, beyond a consciousness of mind around where the items come from. How can we build in sustainability as a part of the physical material makeup of products, whether that be the one-time use of plastic cups or decommissioning batteries inside mobile phones? How can we return those precious materials of the global commons to supply chains? How do we bring the circular economy into the forefront of measuring GDP rather than just pure exports? How can we measure countries' success based on the circular economy and use of these rare and limited resources that we have?

What was the hardest obstacle you had to overcome?

The fact that as human beings we need at least eight hours of sleep.

What was your big breakthrough?

That's like asking an Olympic gold medalist, "What was your big breakthrough moment?" There were ten years of background persistence, training, dedication, and physical and mental fitness to enable you to be ready to execute at a certain point in time. Nothing miraculously happens overnight. It's a patchwork quilt of experience within certain market conditions and the maturity of technologies, all combining and converging at a certain point in time. This has been [the result of] experience and persistence in an industry, combined with that industry's own challenges and an antenna sense—not necessarily a sixth sense but an antenna sense, like an insect that can understand changing weather patterns.

Why should younger women consider STEM and/or entrepreneurship?

With the Fourth Industrial Revolution and the combination of exponential technologies coming together—whether that be the way we handle data security, the application of robotics, or the introduction of artificial intelligence—it's clear that process-driven roles within manufacturing might very well change the skills landscape. Everybody should be thinking about how they can increase their digital intelligence. We have striven to speak multiple [human] languages so that we can communicate. The same thing needs to be at the forefront of our thinking about a digital landscape. How can we learn some form of digital intelligence or even a digital literacy?

When you think about the mechanisms of technology enabling linear processes, then, clearly, we need to advance emotional intelligence. How can we bring ourselves into more of an artistry role? How do we imagine the future-possible, create the tools—and then allow the tools to create us?

Entrepreneurship is just another form of creativity. For me, instead of having a blank canvas with a set of paints and a paintbrush, I use digital crayons and a digital canvas to create my artistry. The process-driven revolution now going on means the machines will take away some of the hard lifting. Then we will all move over into some form of artistry.

Allison Clift-Jennings / Filament

Allison Clift-Jennings was born in Reno and received a BS in computer science from the University of Nevada, Reno. She has been in the technology industry for more than two decades, and has worked in a variety of jobs, from junior database administrator up to CTO. She is currently the CEO of Filament, a tech startup dedicated to a fully decentralized Internet of Things ecosystem that operates independently of central authority. Filament's devices rely on a blockchain-based payment system and digital smart contracts and are, unlike the cloud, designed to operate independently of existing cellular and Wi-Fi networks.

What is your background?

I've worked in the technology industry for twenty-two years, holding roles from junior database administrator up through CTO. I currently hold the CEO role for Filament, a tech startup focusing on bringing trust to the mobility/automotive industry.

Why did you start your company?

I've always been a big believer that technology has the ability to solve many problems. Certainly not all problems, and it sometimes creates new ones—but, fundamentally, technology as a tool can overcome significant barriers in many areas of our lives.

Initially, it was perhaps a bit naïve. [I felt that] if you want something done differently, just go build it yourself. While that ideal has led to quite a few sobering realizations, it also showed me that there is much less holding you back from realizing a future than you think. Almost all of the holdback comes from within yourself. If you get past that, the rest is manageable.

How did you get your funding at each stage?

For Filament, we've pursued funding through traditional venture capital. VC gets a bad rap, and while there are indeed bad actors in any group of people, we've had a very positive experience overall in regard to finding the right investors who share our vision of a trusted vehicle platform.

What advice would you give to other women in your field?

It's so very important to take the risks necessary to move closer to your future vision. It sounds somewhat cliché, but it is absolutely true. Usually, most of the fear you may hold in pursuing that job, or starting that company, or quitting that relationship doesn't match the reality. In many situations, once you move through it and see that the fear was minor, you may be upset with yourself for believing the fear at all. Once you do that a few times, you start to take fear—a natural emotion—and contain it appropriately, as just one voice in a many-voiced group guiding you forward.

What advice would you give to STEM students?

STEM contains a large set of disciplines. If a student is interested in STEM, I would encourage them to explore all areas of science and engineering—even areas they may not be interested in initially. As an example, I started my engineering studies in mechanical, then moved to electrical, and then on to computer science. They all were interesting in their own ways, and I find some things I learned in one discipline actually helped me understand other disciplines better. Don't forget bioscience and formal logic either—also a part of STEM!

What do you predict will happen in your field in the next five years?

The well-known computer scientist Alan Kay is quoted as saying, "The best way to predict the future is to invent it."

In that vein, I spend most of my time focusing on building the possible future that we envision. Will it be *the* future? I don't know, but it's a possibility. And that possible future includes machines, vehicles, and infrastructure we rely on today, but has the ability to become transactive in nature—where machines are economic, and can establish proof and value directly with each other, and with people. If you consider what e-commerce did for the early internet, consider what machine-commerce will do for the industries we rely upon today.

What was the hardest obstacle you had to overcome?

A few years ago, I was in the middle of a fundraising round for Filament. Those familiar with technology start-ups know that fundraising is like a second full-time job for the CEO, [companion] to the primary job of running a company. But what very few people knew outside of my team and my family was that I was in the middle of a gender transition at the same time. Identifying as a transgender individual during the fundraising process brought a large amount of anxiety to me, as I didn't want to bet the future of our company on the conscious and unconscious biases of interested investors. This obstacle led me to some really difficult times, but wonderfully, every single investor was incredibly supportive once I came out to them. One even stated that had I let them know earlier, they could have got the deal done faster due to pulling from a diversity fund. This is, yet again, an example of a fear being incommensurate to the actual events that happen. Fear controlling us is a dangerous inhibitor to living our fullest lives.

What was your big breakthrough?

Early in the history of Filament, we were in a stage as a company where we needed to find a new way to provide our capabilities to customers—often very large industrial companies. Unlike consumers, large corporations often prefer to purchase products as an operational expense, meaning they don't actually own the devices; they simply pay for their use. But we had a problem where we needed to provide this operations expense—or OpEx—to them, but on physical hardware. This would be easy to do if our devices were always connected to the internet, but they often are not. Our big breakthrough was going deep into our engineering roots to invent a way for a machine to be paid for per use, on physical hardware, online or offline. This, to our knowledge, had never been accomplished before, and helped us to build some larger customer relationships that we still have today. Sometimes, raw engineering is necessary to invent capabilities that simply didn't exist prior. This, to me, is the beauty of STEM.

Why should younger women consider STEM and/or entrepreneurship as a career option?

Because women have so much to contribute to these fields! I'm absolutely of the belief that if you have an interest in a field, you should pursue it as far as you want. It's an unfortunate truth that inherent bias prevents women from doing this as easily as other demographics. Yet we know of study after study that shows more diverse teams being more effective.[*] It's important to note that diversity goes further than gender or race. It also includes worldviews, prior experiences, inclinations, interests, and the like. In my opinion, we should all strive to be as diverse as possible along as many axes as possible, as that will make us more effective in almost every way. Women can reclaim the ability to excel with as little inhibition as possible. Pursuing a career in these fields is one of the best ways to weaken that bias.

---

[*] David Rock, Heidi Grant, and Jacqui Grey, "Diverse Teams Feel Less Comfortable—and That's Why They Perform Better," *Harvard Business Review*, September 22, 2016, https://hbr.org/2016/09/diverse-teams-feel-less-comfortable-and-thats-why-they-perform-better.

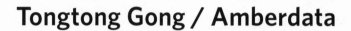

Tongtong Gong / Amberdata

Tongtong Gong is cofounder and COO at Amberdata, a platform for monitoring, searching, and analyzing public and private blockchain. Before becoming an entrepreneur, she was a successful executive and team leader. She served as vice president of engineering at Unified. At Acxiom, she held a variety of engineering and leadership roles and was responsible for establishing a Global Service Center for the company in China. Additionally, she led an innovation and advanced development team for the CTO office at EMC/Dell.

What is your background?

I grew up in Beijing. I came to the United States when I was seventeen years old through an exchange program where you get to live with an American host family and go to school at the same time. Then I stayed for college and work.

I majored in computer science because my English wasn't

good enough to write papers. That was my weak spot, and I felt very vulnerable about it. I didn't understand that my English would eventually be good enough. I thought it would always and forever be a challenge of mine. So I asked my advisor, "How can I survive college without writing papers? What's the major with the fewest papers, the fewest English requirements?" My advisor said, "Try a computer science course. You'll learn a programming language along with everybody else." I figured, "Hey, if you put me on the same starting line with everyone, I can do this. I'm not at a disadvantage." So that's how I started in computer science.

I was a software engineer and then an engineering manager until I cofounded Amberdata last year. I'm a COO now only because I'm lucky enough to have two other cofounders who are equally super-technical. We figured that, out of the three of us, I can do what they do, but they can't do what I do, which is stay organized, plan, and keep things running—make sure everything's taken care.

Why did you start your company?

Around 2013, I was offered a job related to Bitcoin. At that point I thought, *Bitcoin is a fraud. I don't even know what it is. It's maybe something to do with people playing video games.* I was so ignorant, and I didn't even try to google it. I just completely ignored it. Then fast-forward to 2016 and 2017, I start looking into the blockchain and cryptocurrency space again, and this time I actually dug in. I started reading the white papers, understanding the technology behind it, and seeing the applications and the potential. I remember telling myself, "This is so interesting. I need to learn about this and be involved with it. I don't care how; I need to work in this space."

Where does your interest in building companies and products come from?

Three years ago, I joined a startup because I felt like I had hit the glass ceiling working for big corporations. I felt like I could do more. I didn't enjoy the corporate red tape and layers and layers of approval and all the politics. So I joined a startup, and I really, really enjoyed working there because it gave me a lot of freedom and a lot of responsibilities. Things moved fast, so I was able to tell whether my decision or contribution was successful or a mistake and course-correct fairly quickly. I worked for that startup for two years. I learned so much, and I loved the culture.

But then, again, I felt like I could still do more. I wanted to start something on my own, if I could find the right partner and the right idea. That seed was planted and became a voice in the back of my head. Working for a startup brought out that motivation in me. I never thought

I would be an entrepreneur. Growing up in China, the culture was all about staying within the box—be a good student, listen to the teacher, be obedient, follow the rules. I always thought I would be a great employee that would make my boss happy. That's what I was good at: I was accountable; I was responsible. I enjoyed working for the startup so much, though, and I figured cofounding something, if the opportunity arose, would be even better. The little entrepreneurial bug had bitten me, and that led to my growth.

How did you get your funding at each stage?

When we started the project, my two cofounders and I bootstrapped and put the money in ourselves. We didn't take salary, to make sure we could pay for the basic stuff with the money we'd put together. We wanted to build something and then raise money.

We started building, and then we started socializing with VCs to get guidance and feedback. One of my cofounders used to be a VC himself, and he's pretty well connected. We weren't fundraising at first, just having preliminary conversations about the idea with different firms: "Here's our idea. We're building a prototype. Here are our thoughts about the market, and here's our plan for the business model. Are we off? What do you think? Any suggestions? Are we crazy? What should we watch out for?"

One of the investors we asked for advice ended up offering us a term sheet. We were very lucky to get our seed round funding fairly easily. We'd begun building the prototype, but we really didn't have anything but an idea yet. That idea resonated, though, and our background was relevant and positioned us to be successful. We weren't doing something brand new; we were applying what we already knew to blockchain.

What advice would you give to other women in your field?

Don't feel like you *have* to start a company or *have* to do *x*, *y*, or *z* to be successful. Looking back, I think every project I've worked on, every book I've read, every meetup I've gone to, and every person I've met were all preparing me for today, but being a founder isn't for everyone. It's time consuming. It's difficult. It's stressful. I'm fortunate to have a super-supportive husband, but it's still a lot. It's a 24/7 job, and not everyone will enjoy it. Whatever opportunity you have in hand, make the best of it. Whether you work for a big corporation or a smaller company, just do a good job at whatever you do and see where that step takes you next.

As an immigrant myself, I would say to other immigrants: don't feel like you're a failure if you can't start your own company for visa reasons. For the longest time, I couldn't start my own company because I was an H-1B employee on a work visa.

Also, don't feel like you're taking a job from someone else. I was recently at a dinner with a bunch of awesome women, and we were all asked to share and explain a number that was special to us. I chose seven, because that's how many jobs I've created with my company. I'm really proud of that because people talk about how H-1Bs are taking away jobs from Americans. People might say that I took one job away back in the day, but now I have created seven, right? And hopefully that number will grow over time.

What advice would you give to STEM students?

Go where your interest is. Keep learning new things and keep having fun with it. I honestly always thought I'd be a writer, because I loved reading novels and was a good writer, until I had to write in a different language. In a way, I accidentally became a computer engineer, but I really enjoy this field because it's new and there is so much going on. You can always learn something different. Just keep the interest level high.

What do you predict will happen in your field in the next five years?

In blockchain, five years is ten years. Blockchain has the potential to change the way we live our lives and to change our expectations for how technology is used.

Today we subconsciously sacrifice a lot of things, like trust and privacy, for the sake of convenience. If you want to keep up with your friends, you use Facebook, and you tolerate the ads and tolerate your information being sold. If you want to search for something on the internet, you use Google, and you put up with Google deciding the rankings, displaying ads, and knowing everything you ever searched for. We put up with a lot of things because we want the convenience.

Blockchain has the potential to give back what we've given away. We're going to get there, but right now, it's still early. There are a lot of protocol-level things that have to be worked out. That's why I'm so passionate about building what I build. Amberdata is an enabling ecosystem that helps the adoption of blockchain. We provide operational monitoring, exploration, search, analysis, dashboarding, and access to data for blockchain. Essentially, our platform lets you see what's going on with the blockchain—the data, the transactions, the holders, and the addresses. We built this platform so that people can understand blockchain. I believe you need that understanding for users to stop being intimidated by blockchain, for developers to better develop and run on blockchain, and for investors to have confidence to invest more. Blockchain is still very opaque to people today, so we are working on establishing trust and transparency so that people can feel more comfortable with this technology.

What was the hardest obstacle you had to overcome?

Talking to people. When I was an engineer, all day long I just talked to other engineers. I didn't need to go to networking events, introduce myself to strangers, and do a pitch. For me that was terrifying. A year ago, at a networking event, I might have gone up to someone, said hi, talked about the weather or something, and then wanted to leave. I never thought I'd be able to just talk to strangers, but now I can't shut up about what I do all day long, because I'm so passionate about it. I just want to tell everyone about blockchain and what I'm building.

What was your big breakthrough?

My big breakthrough came after we launched the platform. We formed the company in August 2017, launched the beta in February 2018, and launched the production platform in May. It moved really quickly. Up until May it felt like just an idea. We didn't know if people were going to use it or think it was useful.

After we launched it, I was at an event and happened to be wearing an Amberdata T-shirt. A guy walked up to me and said, "Oh, you work for Amberdata? You guys are awesome." I was surprised to be recognized. "You know who we are?" I asked, and he said, "Yeah, Vitalik said you guys are doing an awesome platform." Vitalik is the creator of Ethereum, and he is like a god in the crypto-space. "Vitalik knows who we are? And you know who we are? And you're actually using our platform and think it's awesome?" That was the moment I realized we were no longer just building something within our heads but building something that people cared about and that really mattered.

Why should younger women consider STEM and/or entrepreneurship as a career option?

I love to mentor and encourage people, but I don't think women have to be in STEM. You can excel in any field. Whatever floats your boat, whatever you feel you're good at—do it. Of course, if you choose STEM, all the power to you. I do believe financial independence is really important and having a STEM degree may help you achieve that independence easier than having an art degree.

Is there anything else you'd like people to know about you?

I don't know where the blockchain and cryptocurrency space is going, but I want to be part of it. I just want to help in any way I can to live in a better world.

Rachel Thomas / fast.ai

Rachel Thomas is a cofounder of fast.ai, a deep learning research lab that created the Practical Deep Learning for Coders course taken by over two hundred thousand students. Her writing has been translated into Chinese, Spanish, Korean, and Portuguese, and read by a million people. An expert in AI, she is a frequent presenter and keynote speaker. In 2017, *Forbes* magazine named her to a list of twenty-plus "Incredible Women Advancing A.I. Research." She earned a PhD in math at Duke University. She was an early engineer for Uber and is currently a professor at the University of San Francisco.

What is your background?

I grew up on the Gulf Coast of Texas, not in a major city. I attended a poor, public high school that was later ranked in the bottom 2 percent of Texas schools. I could see how my school had far fewer resources than wealthier schools, and that we were treated differently. This is part of why I care so much about inequality in education and access.

Despite this, I was lucky that I was able to take two years of C++ programming in high school (it still feels like a fluke that I had this opportunity!). My teacher was a fantastic woman, a former programmer who inspired me. She retired right after I graduated and was never replaced, so my younger brother didn't get to take computer science at all.

I majored in math, with minors in computer science and linguistics, at Swarthmore College, then earned a PhD in math at Duke University. My original goal was to become a math professor, although I was discouraged by sexism and harassment that I experienced in grad school, as well as ways that higher education is changing (universities laying off tenured faculty during the financial crash, friends doing multiple postdocs before finding a permanent position if they are lucky, etc.).

After finishing my PhD, I worked as a quant in energy trading, which is where I first started working with data. In late 2011, a friend sent me an article about "data scientist" being an exciting new career. I applied to a few of what were my "dream jobs" at the time and was rejected. I accepted an analyst position at a startup I was excited about, with the hope and an informal agreement that I could move into an official data science/modeling role later.

I didn't know anyone who worked in tech and had never

lived on the West Coast, yet I got rid of my car, furniture, and most of what I owned to move to San Francisco with just two suitcases and my new dream of becoming a data scientist. My first year in San Francisco was a period of intense learning: I attended tons of meetups, completed several online courses, participated in numerous work-shops and conferences, learned a lot by working at a data-focused startup, and most importantly, met scores of people whom I was able to ask questions of. I completely underestimated how amazing it is to be able to interact regularly with the people who are building the tools and technology that excite me most.

I landed a job as a software engineer and data scientist at an up-and-coming startup. Although I liked the work, I found the environment in tech to be sexist, isolating, and aggressive. It was devastating to have worked so hard for a goal and then discover that I was miserable. I even considered leaving the tech industry altogether at one point and hired a career counselor to explore my options. I eventually returned to teaching (which I love), and then later started fast.ai.

Why did you start your company?

I could see the huge potential of deep learning (a high-impact field of AI), as well as how hard the field was to get into and to use. Our Practical Deep Learning for Coders course is the resource that I wish had existed five years ago when I was first getting interested in deep learning.

When I became interested in deep learning in 2013, I found that experts weren't writing down the practical methods they used to actually get their research to work and were instead just publishing the theory. I believe deep learning will have a huge impact across all industries, and I want the creators of this technology to be a more diverse and less exclusive group. People can't address problems that they're not aware of, and with more diverse practitioners, a wider variety of important societal problems will be tackled.

At fast.ai, we are working to make deep learning more accessible through several avenues: doing research and building software to make the technology easier to use, as well as providing practical education with no advanced math prerequisites that takes coders to the state of the art.

I'm interested in making deep learning more accessible, and at the time we started fast.ai, I didn't see anyone making deep learning accessible in the ways that were needed.

Tell me about how you've funded the company.

I'm leery of venture capital, as the incentives of VCs are

often misaligned with the incentives of founders. Most VCs are looking for just a few of their companies to be wildly successful and are not interested in moderate success, so they pressure startups into hyper-growth, which typically is bad for people, bad for products, and ultimately destructive. (I've experienced this at several startups I worked at, and I've also witnessed unethical, bullying, and downright illegal behavior by venture capitalists against close friends of mine who are founders.)

Growing at a slow, sustainable rate helps keep your priorities in order. Funding yourself (through part-time consulting, saving up money in advance, and/or getting a simple product to market quickly) will force you to stay smaller and grow more slowly than VC-funded businesses, but this is good. Staying small keeps you focused on a small number of high-impact features.

Fast.ai is self-funded. Our costs are low as we have just three people, and we also work part-time at the University of San Francisco Data Institute, which has been very supportive of fast.ai. We have a number of students and alumni who contribute to our open source projects and are actively involved.

Despite being a small team, we are having a much bigger impact than I ever expected. In just two years, we've had over two hundred thousand people around the world take our course. Fast.ai students have landed new jobs, launched companies, had their work shown on HBO, been featured in *Forbes*, won hackathons, and been accepted to the Google Brain AI Residency. Fast.ai has been covered in the *MIT Tech Review* and *The Economist* and mentioned in the *Harvard Business Review* and the *New York Times*. A fast.ai team beat engineers from Google and Intel in a competition hosted by Stanford, and this victory was covered in *The Verge* and *MIT Tech Review*.

Having a small team forces us to prioritize ruthlessly, and to focus only on what we value most, or think will be highest impact. Something that has surprised me is how much I've been able to invest in my own career and own skills, in ways that I never could in previous jobs. My cofounder and I are committed to fast.ai for the long term, so neither of us has any interest in burning out. We believe you can have an impact with your work, without destroying your health and relationships. I'd love to see more small companies building useful products in a healthy and sustainable way.

What advice would you give to other women in your field?

For years, I felt like it was a personal failing when I couldn't thrive in sexist or toxic environments, and I believed that I just needed to be smarter, tougher, or more confident. Ultimately, I realized that the problem is with the toxic environments, not with me. The main things I would tell women (and everyone) is to be compassionate with yourself, and to minimize time spent with toxic people and toxic environments as much as you can.

What advice would you give to STEM students?

I used to assume that the confidence of others meant that they knew more than me, but I saw again and again that this was not the case. Throughout my education, I saw guys who were much more confident than me, but they didn't actually understand the material well.

Secondly, learning is not a race. When I tried to take Real Analysis (a course required for all math majors) the first time, I couldn't understand anything and was miserable. I dropped out of the class after a few weeks, and I told everyone that I didn't like Real Analysis and wasn't good at it. A year later, I enrolled in the course again with a different professor. This time I loved it! I earned an A, went on to take more advanced Real Analysis courses, and even chose to focus on the topic during my PhD! Just because you struggle with a course or topic, that doesn't mean that you're "bad" at it or that you won't be able to master that topic if you try again at a different time or in a different context.

Also, I encourage everyone to read the book *A Mind for*

Numbers: How to Excel at Math and Science (Even if You Flunked Algebra), which has lots of great research-backed strategies about effective studying and learning (it is used in the Coursera Learning to Learn course).

What do you predict will happen in your field in the next five years?

Just as all companies now use the internet, in the future all companies will use deep learning. We are just on the cusp of this with deep learning, perhaps where we were in the early '90s with the adoption of the internet.

Deep learning is already being used by fast.ai students and teachers to diagnose cancer, stop deforestation, provide crop insurance to Indian farmers, help Urdu speakers in Pakistan, develop wearable devices for patients with Parkinson's disease, and much more. Deep learning offers hope of a way for us to fill the global shortage of doctors, providing more accurate medical diagnoses and potentially saving millions of lives. It could improve energy efficiency, increase farm yields, reduce pesticide use, and more.

There's a lot of potential for both good and bad. I'm excited about the ways that deep learning can improve medicine, agriculture, and the environment, although I'm scared about how it is being used for surveillance, promoting conspiracies, and encoding unjust biases. I think it is crucial to get a more diverse group of people involved, both to address the misuses and to take full advantage of the positive opportunities.

What was the hardest obstacle you had to overcome?

My lowest points occurred when I was dealing with sexism, harassment, and generally toxic environments. This happened in graduate school, and then several years later in the tech industry. In both cases, I had spent years making sacrifices and working toward a goal (in the first case to be a math professor, and then later to become a data scientist). It was devastating to put that much into my goal and then end up depressed, burnt out, and unsure of what to do next.

Toxic environments cause you to doubt your own perceptions, skew your sense of what's normal, and make you feel like your unhappiness is your own fault.

What was your big breakthrough?

I don't know that I've had just one big breakthrough—my career has had a number of ups and downs. Depending on the point in time you talked to me, there were instances when everything felt like a failure, but now when I look back, I can see that all my experiences have helped me toward the work I'm doing now.

Why should younger women consider STEM and/or entrepreneurship as a career option?

I hope that women (or anyone facing a toxic work environment) will consider entrepreneurship or working for themselves. It can be a way to have more control over the work you do and the environment you are in. I know a few women who have been so fed up with their jobs at tech companies that they've started consulting for themselves and really enjoyed the increased autonomy.

STEM is a high-impact field with lots of exciting developments happening. It gives you an opportunity to be creative, to build powerful tools and inventions, and to address a range of problems. I would encourage everyone to have some understanding of *both* sciences/tech *and* the social sciences/humanities, because we are seeing so many problems that lie at their intersection.

Conclusion

The stories in this book are evidence that women can conquer hard technologies. Entrepreneurship is hard for everyone, women and men alike; the people who succeed are the ones who are able to persevere in the face of difficulty. Obviously, you need a certain amount of confidence to do that, but beyond that, what I see in the entrepreneurial women I know is passion.

Take Carrie Shaw: she cares deeply about the product she's offering to the elderly and their caregivers. She's not going to give up or even be too discouraged day-to-day, because she knows that she's in the right place, doing what she needs to be doing.

Another theme that I hope you drew from the stories you've read here is the central role technology and

STEM will play in the future. Both Olga Egorsheva and Breanna Faye married an interest in art and design with technology to create their companies. Neither is at root a tech person (Olga studied business, and Breanna studied architecture), but both are working in tech and are excited about its possibilities. Most of the women you've read about believe no one should feel excluded from the world of tech-based startups because they lack a STEM background or education.

Nor do you have to be a trained entrepreneur in order to start your own company. Tongtong Gong, for example, never would have guessed she'd be an entrepreneur. She was an engineer with a really strong background in tech, not business. But she had an idea that she wanted to pursue as the founder of a startup. She's thriving in her new role, and she sees how being an entrepreneur has helped her get her ideas out into the world in a way that probably wouldn't have happened if she had continued to work for someone else.

If you are an investor, I hope what you have read excites you as it does me. I have offered a glimpse in these pages of what women are doing on the cutting edge of technological innovation. I invite you to approach their work, and the work of thousands of other entrepreneurial women, with an open mind, and to consider whether you can make money with them (hint: you can!).

If you have your own experiences regarding women in technology—as an investor, a founder, or simply someone who wants to get involved—I hope you'll share your stories with me. I'd be delighted to hear from you.

Acknowledgments

I would like to thank my family for their support while giving up weekend time with them. I hope this inspires my daughter to be interested in STEM. To my parents, for their unwavering support.

I would also like to thank Rachel King and Wendi Goodman for their support and editing. Thanks to Craig Enenstein, whom I met through the Wharton network, for an impassioned discussion that led to his thoughtful foreword.

To all of the female entrepreneurs and investors who impress and inspire me every day, you know who you are—thank you infinitely.

About the Author

NISA AMOILS is a venture capital partner at Republic Labs, focused on disruptive technologies. She is a former securities lawyer and on the boards of several companies, Girls Who Invest, and Wharton Entrepreneurship. She writes for *Forbes* and *Blockchain Magazine* and is a regular judge/panelist on CNBC, MSNBC, Fox, Cheddar, and others. She has been involved in many startups and spent many years in business development at Time Warner and NBC Universal. She holds a business degree from the University of Michigan and a law degree from the University of Pennsylvania. www.nisaamoils.com